Hard Road West

HARD ROAD WEST

HISTORY & GEOLOGY ALONG
THE GOLD RUSH TRAIL

* Keith Heyer Meldahl *

The University of Chicago Press   CHICAGO & LONDON

KEITH HEYER MELDAHL is a professor of geology
and oceanography at Mira Costa College in Oceanside, California.

The University of Chicago Press, Chicago 60637
The University of Chicago Press, Ltd., London
© 2007 by The University of Chicago
All rights reserved. Published 2007
Printed in the United States of America

16 15 14 13 12 11 10 09 08 07   2 3 4 5

ISBN-13: 978-0-226-51960-9   (cloth)
ISBN-10: 0-226-51960-0   (cloth)

Library of Congress Cataloging-in-Publication Data

Meldahl, Keith Heyer.
Hard road west: history and geology along the Gold Rush trail / Keith Heyer Meldahl.
p.   cm.
Includes bibliographical references and index.
ISBN-13: 978-0-226-51960-9 (cloth: alk. paper)
ISBN-10: 0-226-51960-0 (cloth: alk. paper)
1. Pioneers—West (U.S.)—History—19th century. 2. Pioneers—West (U.S.)—Biography.
3. Frontier and pioneer life—West (U.S.). 4. Overland journeys to the Pacific. 5. West
(U.S.)—History—1848-1860. 6. West (U.S.)—Description and travel. 7. West (U.S.)—
Geography. 8. Landscape—West (U.S.)—History—19th century. 9. Landforms—West
(U.S.). 10. Geology—West (U.S.). I. Title.
F593. M479 2007
978'.02—dc22
2007011052

To my mother, who always wants to know more
And to my father, who celebrates life

History is all explained by geography.

ROBERT PENN WARREN

Any man who makes a trip by land to California
deserves to find a fortune.

ALONZO DELANO, 1849 emigrant

One only hope sustains all these unhappy pilgrims,
that they will be able to get to California alive,
where they can take a rest, and where the gold
which they feel sure of finding will repay them for
all their hardship and suffering.

MARGARET FRINK, 1850 emigrant

CONTENTS

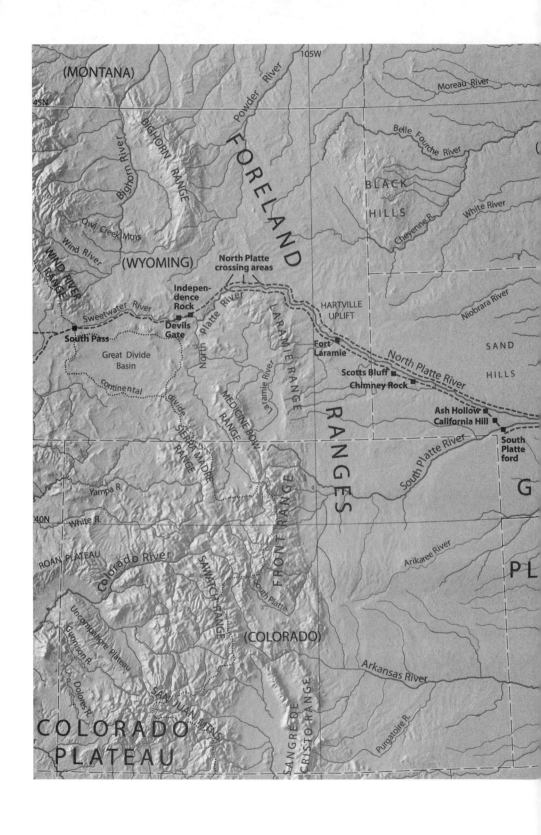

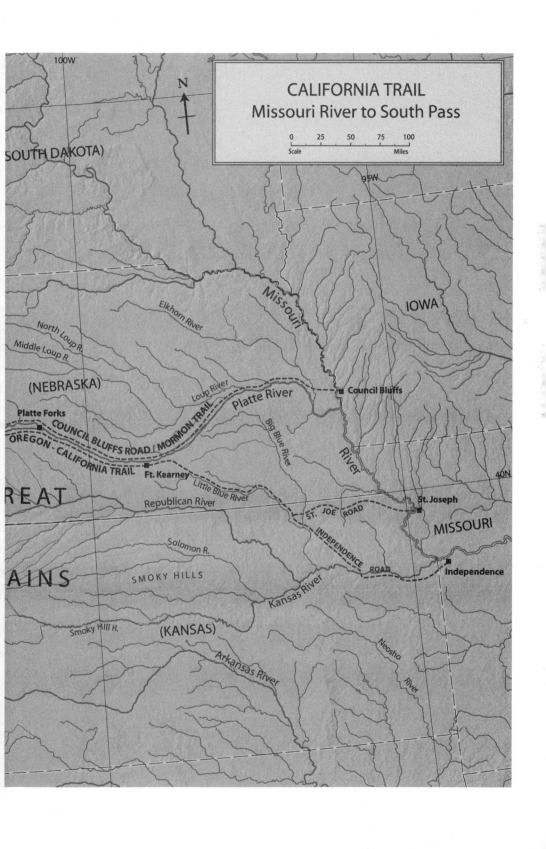

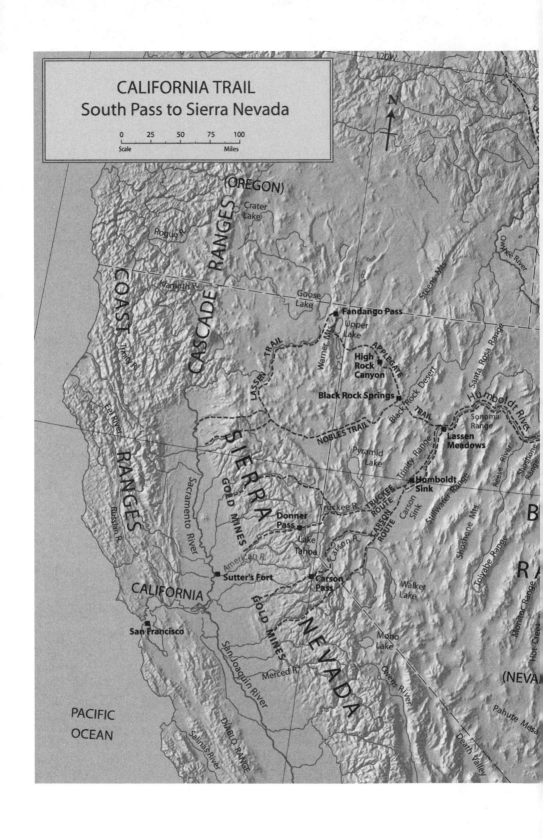

CALIFORNIA TRAIL
South Pass to Sierra Nevada

Scale 0 25 50 75 100 Miles

120°W

N

(OREGON)

Crater Lake

Rogue R.

COAST RANGES

CASCADE RANGES

Trinity R.

Klamath R.

Goose Lake

Fandango Pass

Upper Lake

Warner Mts.

High Rock Canyon

APPLEGATE

Black Rock Springs

Black Rock Desert

BLACK ROCK TRAIL

Humboldt River

Santa Rosa Range

Sonoma Range

LASSEN TRAIL

NOBLES TRAIL

Lassen Meadows

Pyramid Lake

Trinity Range

Reese River

Shoshone Range

Feather River

SIERRA

GOLD MINES

Sacramento River

Humboldt Sink

Truckee R.

TRUCKEE ROUTE

CARSON ROUTE

Carson Sink

Stillwater Range

Shoshone Mts.

B

Donner Pass

Lake Tahoe

Carson R.

Towabe Range

R A

American R.

Sutter's Fort

Russian R.

CALIFORNIA

Carson Pass

GOLD MINES

Walker Lake

Monitor Range

Hot Creek

San Francisco

NEVADA

Mono Lake

(NEVA

PACIFIC OCEAN

San Joaquin River

Merced R.

DIABLO RANGE

Salinas River

Owens River

Death Valley

Pahute Mesa

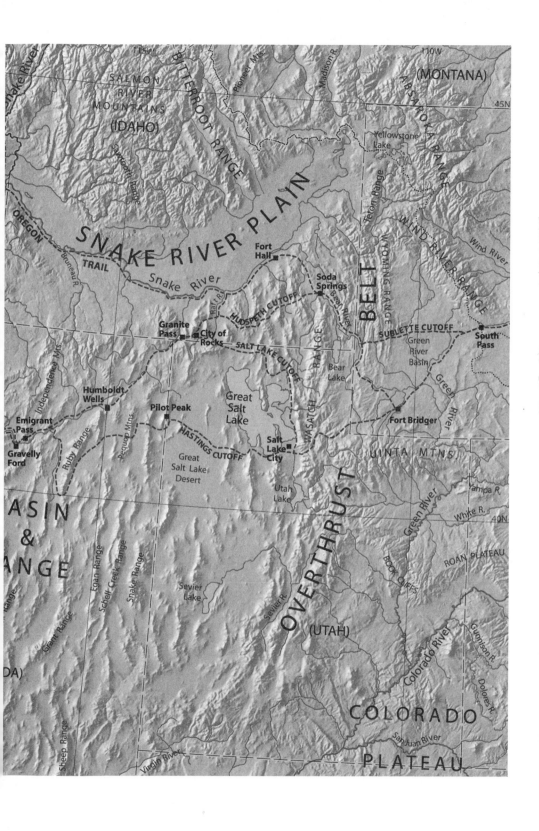

Geologic Time Scale

EON	ERA	PERIOD		EPOCH	AGE *	Events in western North America
Phanerozoic	Cenozoic	Quaternary		Holocene	0.01	
				Pleistocene	1.8	
		Tertiary	Neogene	Pliocene	5.3	
				Miocene	24	
			Paleogene	Oligocene	34	
				Eocene	55	
				Paleocene	65	
	Mesozoic	Cretaceous			142	
		Jurassic			202	
		Triassic			250	
	Paleozoic	Permian			292	
		Pennsylvanian (Carboniferous)			320	
		Mississippian (Carboniferous)			354	
		Devonian			417	
		Silurian			440	
		Ordovician			495	
		Cambrian			545	
Precambrian		Proterozoic			2500	
		Archean			3800	
		Hadean			4550	

Events in western North America (right column, keyed to ages):

- "Ice Ages" — Yellowstone – Snake River caldera tract (still active) — Exhumation of the Rocky Mountains (ongoing) — Basin & Range Orogeny (ongoing)
- Caldera eruptions across the future Basin & Range
- Lake Gosiute (WY); auriferous gravels (CA)
- Sevier Orogeny — Laramide Orogeny
- terrane accretion assembles much of California — Nevadan Orogeny
- Pangea starts to break up
- North America merges with other continents to form supercontinent Pangea; terranes accrete in Nevada — Thick stacks of strata build up on the Great Plains
- Core of North America assembled by serial collisions of crustal blocks — Great Plains Orogeny adds more land to North America
- Earth forms

* In millions of years before present. Note that the vertical scale varies; older geologic intervals cover greater time spans than do younger intervals.

PREFACE

Several years ago, while exploring in western Nevada, I found myself at the Humboldt Sink, a dust-caked playa (dry lake) where the Humboldt River dies, sucked up by thirsty ground and dry air. It was August, blazing hot and bone-dry. The sun pushed down like a heavy weight. A sharp wind blew dust devils about and filled the air with grit. No living thing appeared on the five-mile-wide, utterly flat, cracked mud-and-salt surface of the sink. Bare, brown mountains rose up all around: the West Humboldt Range to the east, the Trinity Range to the west. From this distance their gullies and ridges, shimmering with heat, were unsoftened by any hint of vegetation. Plants do live there—shrunken miserable things that hug the ground, waving nasty spikes, waiting for winter rain. This is a world ruled by rock, dust, and heat, where the land swallows rivers and life peeks from the corners. It is about as surly as nature can be.

I found there a solitary post marked "California Trail"—the overland route used by California-bound emigrants during the gold rush years. Faint parallel grooves in the desert floor stretched a few yards and then died in the greasewood shrubs. Wagon ruts. Bits of detritus lay beside the post—a rusted fragment of a wagon-wheel rim, pieces of barrel hoops, broken bits of crockery. Historic trail buffs (or "rut nuts," as they call themselves) had gathered these up from the desert, where they had laid since about 1849, and placed them respectfully by the post. I crouched and picked them up. What an ordeal, I thought. Imagine being stuck out here with nothing but a tattered wagon for shelter, creaking along toward California at a walking pace—15 or so miles per day.

What tough people.

What an adventure.

I cracked a cold beer in their honor. (Never head into the field unprepared.) The seeds of obsession sprouted in me on that desiccated day.

I have spent much of my life exploring the West and teaching its geology to my students. For 20 years the rough land has lured me with its wild beauty, naked mountains, and stunning rock exposures. In the humid East, a cloak of annoying vegetation covers the rocks, and geologists line up to hammer at road-cuts—precious gifts of highway construction. In the arid West, the Earth leaves that prudish green mantle behind and bares her skin, etched with stories from the depths of time.

But emigrant journals from the gold rush era showed me a different West. This West was a brooding menace—a fearsome land of sharp rock and little water, a vast and pitiless place where each day brought new hardships. This West tested all who dared cross it and defeated many. "[I have] undergone more hardship than I ever thought possible to live through," William Wells wrote as he plodded toward California in 1849. Swept up by emigrant stories, I became a rut nut. Emigrant diaries, trail history books, and maps piled up around the house, festooned with markers and scribbled notes. Geologic maps and emigrant diaries in hand, I eventually chased emigrant shadows for 2,000 miles over the Great Plains, through the Rocky Mountains, and across the Great Basin to the Sierra Nevada and the gold fields—to the Mother Lode.

The more I explored, the more I saw how North America's story—the history of a continent written in its rocks—set the course of America's story. For millions of years, our continent has migrated west. It left Eurasia and Africa behind, opening the Atlantic Ocean in its wake and overriding thousands of miles of ancient seafloor along its western edge. That singular history assembled and carved the landscape through which the emigrants passed—a stunning landscape that evoked their awe and wonder even as they cursed its hardships. That history set the golden bait that lured thousands to California with dreams of wealth. It also put a brutal gauntlet of mountain barriers in their way. That history decreed the courses of rivers—those life-sustaining corridors that determined the routes of the overland trails. It raised mountains that choked moisture from western skies, resulting in deserts that exacted tribute in the form of thirst, hunger, abandonment, and death. In short, North America's geologic story built the stage and the props, and wrote large parts of the script, for the human drama of the westward migration. This book tells both of these stories—one, the story of the overland emigrant journey, and the other, the tale of the land itself: of rocks, rivers, mountains, and deserts, and how they came to be.

THE WESTWARD migration of the mid-nineteenth century was the greatest mass migration in American history. More than 400,000 souls passed overland to Oregon, California, and the Salt Lake Valley between 1841, when the first overland emigrants trickled into Oregon and California, and 1869, when the completion of the transcontinental railroad virtually ended wagon-train emigration. The peak of the emigration was to California from 1849 to 1853—the first five years of the gold rush. The emigration forever transformed the face of the nation and, for better or worse, fulfilled the prophecy of Manifest Destiny—the notion that Euro-Americans from the East had a God-given right to spread west and establish dominion over the continent.

The rocks and mountains of the West have changed little since 1849 (unlike most of the native animal and plant populations and native cultures). The geologic landscape along the trails needs no reenactments, no props, no tricks of animation to re-create historic authenticity. It is genuine. Subtract the buildings, highways, and reservoirs, and you see the landscape much as the emigrants saw it. The past becomes personal when you stand in the old wagon ruts and read what emigrant men and women thought and wrote while looking out at the same scenes. Byron McKinstry, an 1850 pioneer who kept his diary going even through the worst of times, was once chided by a companion, "My God, McKinstry, why do you write about this trip? All I hope for is to get home, alive, as soon as possible, so that I can forget it!" Luckily for us, there were many McKinstrys on the road west—emigrants who took the time to write, nearly every day, through the months of toil. Their accounts of the land and their experiences of the journey form the narrative core of this book. I let them tell their own stories as much as possible, and I have not changed spelling or grammar in quoted emigrant accounts.

A geology book for general-interest readers that crosses 2,000 miles of some of the best-studied geology on Earth must necessarily leave many things out. I focus on the big picture—the overarching processes that have shaped the American West over millions of years of geologic time to produce the sprawling, spectacular landscape that we see today. These processes include the piecing together of the continent from once-isolated crustal fragments, the burial and exhumation of the Rocky Mountains, the crustal upheavals that built the North American Cordillera (the mountain belt that stretches from the Rockies to the California coast), the cataclysmic history of the Snake River Plain and Yellowstone calderas, the breaking up of the Southwest to make the Basin and Range Province and the Great Basin, the uplift of the Sierra Nevada, and—at the end of the rainbow—the formation of California's gold. These big themes, as well as many smaller geologic stories and asides, form the scientific core of the book.

"Eastward I go only by force; but westward I go free," Thoreau wrote in 1862. He meant the direction he liked to stroll from his Massachusetts house, but he had a larger point about the West. "I must walk toward Oregon, and not toward Europe," he wrote, for "that is the way the nation is moving." Thoreau knew the pull of the West and understood the rich blood of promise it injected into the veins of the young nation. The nation has changed, but the lure of the West remains.

✳ Introduction ✳

STARDUST

The whole country from San Francisco to Los Angeles, and from
the sea shore to the base of the Sierra Nevadas, resounds with the
sordid cry of "gold! gold, GOLD!" while the field is left half planted,
the house half built, and everything neglected but the manufacture
of shovels and pickaxes.

SAN FRANCISCO CALIFORNIAN, May 29, 1848

We live and die, but we are made of immortal stuff. The carbon atoms
in our fingernails, the calcium atoms in our bones, the iron atoms in our
blood — all of the countless trillions of atoms of which we are made — are
ancient objects. They existed before us, before the Earth itself, in fact.
And after each of us dies, they will depart from our bodies and do other
things. Forever.

Iron atoms are circulating in your blood right now, attached to hemo-
globin molecules and carrying oxygen. They got there because you ate
something that had iron in it, maybe a steak. Before those iron atoms
became part of that cow, they were in the grass that the cow ate. Before
that, they were in the soil, sucked up through the roots of the grass. Before that, they were in the rock that broke down to make the soil. Before
that, the iron atoms may have been sloshing around in a scalding soup of
molten rock deep underground. Before that, they might have been part
of the mud of an ancient seabed that plunged on a moving tectonic plate
into the Earth's hot interior to melt and make the magma from which the
newer rock was born. And so on, back through the abyss of time.

Atoms are the closest things to immortal objects that we know. There
are only about 100 naturally occurring types of atoms on Earth, and we
call them the chemical elements. Most of them existed before there was
an Earth. Where did they come from?

Astronomers think that shortly after the big bang some 12 to 14 billion
years ago, the matter of the universe consisted almost entirely of the two
smallest elements: hydrogen and helium. Heavier elements did not yet

exist. There was not one atom of carbon, oxygen, calcium, sodium, iron, zinc, or any of the other elements that today form grass, granite, cows, conglomerate, babies, basalt, and everything else. But all elements, no matter their size, have fundamentally the same design—a nucleus of protons and neutrons surrounded by halos of electrons. This suggests that larger elements can be made by fusing smaller ones together. But it takes monstrous pressure to do this—pressure far higher than found naturally anywhere on Earth. To make most of the known chemical elements, you need the most stupendous pressures in the known universe. You need a supernova.

As stars like our sun run out of atomic fuel, they swell into red giant stars and then collapse into small white dwarf stars. But larger stars (those eight times more massive than the sun or larger) end their lives in spectacularly violent explosions called supernovae. A star that goes supernova releases so much energy that it may outshine, for a brief time, all of the companion stars in its galaxy. In a single frenzied moment, the pressure blast fuses trillions upon trillions of new chemical elements. The copper in your pipes and wires, the zinc in your vitamin tablets, the lead in your computer's innards—most of the 100-plus known chemical elements—are the offspring of ancient supernovae that exploded long before the Earth existed.

Supernovae have spewed their elemental spawn into space ever since the first stars formed after the big bang. About one supernova happens every second in the universe, perhaps one per century in our galaxy alone. Every supernova enriches its region of the universe with new elements. Clouds of these elements then collapse under gravity to make more stars—and planets too. Exploding stars are both the factories and the distribution centers for the chemical elements.

About 5 billion years ago, in one corner of one galaxy among billions, a cloud of matter, enriched with elements from prior supernovae, contracted under gravity. Most of the material spiraled into the center to form an average-sized star. Enough material was left over, orbiting far out from the star, to form eight planets.[1] Much later, on the third planet out from that star, big-brained bipeds evolved and eventually figured out an astonishing fact: the Earth and its resident life are the coalesced remains of ancient stardust.

One of those bits of stardust sits in row 6, column IB, on the periodic table of the elements, platinum on one side, mercury on the other. It is element 79, chemical symbol Au, from the Latin *aurum*. It is one of the

---

1. The nine planets that most of us learned in elementary school are no more; astronomers recently demoted Pluto from full planetary status.

most inert substances known, emerging shiny and unscathed from attack by a host of caustic agents. A solution of concentrated nitric acid and hydrochloric acid together (neither alone will do it) can dissolve the king of metals, giving the liquid its name: *aqua regia*. It is the most malleable metal known; a piece the size of a marble can be hammered into a sheet the size of a large living room. It is far from the rarest of elements, but it is not common. Its average concentration in Earth's crust is about five parts per billion—more than 6,000 tons of rock for each ounce of element 79. In spite of its rarity, humankind has dug, blasted, and sifted enough of it from the Earth to cover a football field nearly six feet deep.

Element 79 is just one form of stardust among 100 elemental cousins. Yet it has started wars, incited murders, inspired exploration, granted stunning wealth to some, and sentenced many more to poverty. It has turned the course of history. One of those turning points began on a chill winter day in January 1848. The region was called California. The United States would soon take it from Mexico as part of the spoils of war. The place was the western foothills of the Sierra Nevada, along the South Fork of the American River. There, at a large bend in the river, a millwright named James Marshall was building a sawmill in partnership with John Sutter, who owned a trading post known as Sutter's Fort near what is now Sacramento. The bend brought the river back on itself. Marshall knew that if he funneled part of the river through a millrace dug across that bend, he could harness a lot of waterpower. The water would turn a wheel, a saw blade, and hopefully a profit. He and Sutter hired some strong-backed young men recently mustered out of the Mormon Battalion of the U.S. Army. The men were paid 12½ cents for each cubic yard of gravel dug, and they were happy for the work. Once they had dug out the upper millrace and piled the rocks and gravel in the riverbed to make a dam to divert water into the millrace, the efficient Marshall employed the river to do some of the work. The men dug by day, the river by night. Marshall would open the stopper gate to the millrace in the evening and let part of the river pour through to scour the millrace. In the morning he would close the gate so the men could go back to work.

One morning—the date usually cited is January 24, 1848—after closing the gate, Marshall went on his usual inspections. It was cold. Ice crystals flashed from the rocks where the water had recently rushed through the millrace. Near where the millrace rejoined the river downstream, a different flash caught Marshall's eye—a yellowish one. He picked up several pieces of glittering rock about half the size of a pea. "Having some general knowledge of minerals," he later wrote, "I could not call to mind more than two which in any way resembled this—*sulphuret of iron* [pyrite], very bright and brittle; and *gold*, bright, yet malleable; I then tried

Unknown California miner, ca. 1850. Courtesy of Collection Matthew R. Isenburg.

it between two rocks, and found that it could be beaten into a different shape, but not broken." Marshall went over to where the men were working and told them he had found gold. They gathered around in disbelief. One of them hammered a piece into a thin sheet. One of the men's wives, Jennie Wimmer, tossed a sample into a pot of lye, where she was making soap. "If it is gold, it will be gold when it comes out," she reasoned. She soaked the specimen in lye all night, and in the morning "there was my gold as bright as could be." Marshall took some samples to Sutter. Skeptical, Sutter dropped them into nitric acid—to no effect. He pulled his well-worn *Encyclopedia Americana* from the shelf. Consulting the entry on gold, Sutter and Marshall conducted a density test. They balanced equal weights of Marshall's samples with silver coins on an apothecary scale and then immersed the scales in water. The tip of the scales settled any doubt—the sinking gold identified itself with its greater density.

Neither Marshall nor Sutter, nor anyone else, could know how utterly everything was about to change.

As the news leaked out, "it triggered the most astonishing mass movement of peoples since the Crusades," writes the historian H. W. Brands.

People from all corners of the Earth descended on California. From the eastern states alone, about a quarter of a million came overland along the California Trail, scouted and opened just a few years earlier as a variant of the original Oregon Trail. Thousands more came by ship, making the harrowing transit around Cape Horn at the bottom of South America. Thousands of others braved the malarial swamps of Panama to short-cut across the isthmus. Their journeys—by schooner; by steamship; by horse, mule, and ox; by their own feet—were the greatest adventures of their lives. For those making the harrowing overland trek across the American West, it was the hardest thing they had ever done. The journey cleaved each life into *before* and *after*, irreversibly and forever. Many never made it. Those who did got a chance to scratch for gleaming supernova leftovers in the bosom of the Mother Lode.

AN AMERICAN JOURNEY

As when some carcass, hidden in sequestered nook, draws from
every near and distant point myriads of discordant vultures, so
drew these little flakes of gold the voracious sons of men.

HUBERT HOWE BANCROFT, *History of California* (1884)

October was dangerously late to be crossing the Forty-Mile Desert, and
Sarah Royce knew it. Only three years earlier, in 1846, October snows
had doomed the Donner party in the high Sierra Nevada, and the Royces
still had a long pull before reaching those mountains.

But snow was the furthest thing from Sarah's mind right now.
First, they had to cross this desert. The lone wagon made slow progress
through the hammering heat. Sarah, her husband, Josiah, and the three
other adults in the group walked alongside the wagon to spare the oxen,
while the Royce's two-year-old daughter, Mary, rode. Ahead, an ocean of
salt flats and sand dunes stretched to the horizon. Beyond those lay the
Carson River.

Three weeks of hard travel along the salty, foul Humboldt River had
sapped the oxen's strength and nearly finished off their food supplies.
Then the Royces had made a colossal navigation error. Traveling by
night, they had unwittingly missed their last chance to take on water and
grass before heading out onto the Forty-Mile Desert. Many hours later
they had realized their mistake—and what it meant.

Turn back! What a chill the words sent through one. *Turn back*, on a
journey like that; in which every mile had been gained by most earnest
labor, growing more and more intense, until, of late, it had seemed that
the certainty of *advance* with every step, was all that made the next step
possible. And now for miles we were to *go back*. In all that long journey
no steps ever seemed so heavy, so hard to take, as those with which I
turned my back to the sun that afternoon of October 4th, 1849.

The exhausting backtrack had cost precious travel time. Now they again plodded west along the same route.

Knowing she would need her strength for the all-night walk ahead (no one camped in the Forty-Mile Desert if they could help it, and night travel conserved water), Sarah lay down for a nap in the wagon. She woke to her husband's voice, "So you've given out have you Tom?" The ox lay prostrate in the yoke. His partner was also near collapse, unable to pull. They unhitched both animals and left them to die. Four oxen remained to haul the same burden, and the Carson River now seemed even farther from reach. A guilty Sarah resolved to ride no more.

They entered the worst of the crossing as darkness fell. By hazy starlight, they passed through a gauntlet of horrors. Discarded possessions and putrefying carcasses of livestock lined the trail. Abandoned wagons loomed up in the darkness. The owners had loaded what they could onto the backs of their remaining animals and pressed on with no hopes beyond survival itself. These "scenes of ruin . . . kept recurring," Sarah remembered, "till we seemed to be but the last, little, feeble, struggling band at the rear of a routed army." Amid the wreckage lining the trail, Sarah spied a small clothbound book titled *Little Ella*. She pocketed it, thinking it would please Mary. It was a simple gesture of faith—I will read this book to my daughter in better times ahead.

They stopped often throughout the night to rest, eat a little, and feed handfuls of stored grass to the weakening oxen. "So faithful had they been, through so many trying scenes," Sarah reflected, "I pitied them, as I observed how low their heads drooped as they pressed their shoulders so resolutely and yet so wearily against the bows."

The last of the water ran out near dawn, and with the sun arose the understanding that they would not survive the day without water. No one spoke. They trudged on, scanning the horizon in the emerging daylight for some sign of the river.

"Was it a cloud? It was very low at first, and I feared it might evaporate as the sun warmed it." They dared to hope that the smudge Sarah had spotted on the horizon might be timber along the Carson River. The oxen knew before the people could be sure. First one and then another gave a low moan and lifted his head to sniff the wind—with the scent of water and trees that it bore. Salvation. They would reach the Carson River.

TO CROSS the Sierra Nevada, the Royces had to follow the Carson River upstream and then make a steep drive to the mountain crest. With the desert ordeal behind them, the threat of October snows loomed larger in their minds. To be cut off, trapped on the east side of the mountains over the winter in a land with little game, would be to court starvation. There was nothing to do but to press on as fast as possible.

This late in the emigration season, the Royces had no reason to expect company, especially headed east. Yet that is what they saw on October 12, as they rolled west up the valley of the Carson River. Two riders descended toward them out of the mountains ahead. Sarah wrote, "Their rapidity of motion and the steepness of the descent gave a strong impression of coming down from above, and the thought flashed into my mind, 'They look heaven-sent.'"

The riders pulled up. "Well sir," they hailed Josiah, "you are the man we are after!"

"How can that be?" responded Josiah.

"Yes sir, you and your wife, and that little girl, are what brought us as far as this."

The riders were part of a relief party dispatched by the California provisional government to help late-arriving emigrants over the Sierra Nevada. The men had orders to go no farther east than the crest of the mountains; their job was to assist emigrants across the summit passes. But nearly a week earlier, on their forced backtrack in the desert, the Royces had passed another group of emigrants headed west. That group, now several days' travel ahead, had reached one of the summit passes and been immediately trapped in a snowstorm. They had nearly died but had battled their way to the government men's relief camp. There was a woman in that group, and as one of the riders explained:

> [She] set right to work at us fellows to go on over the mountains after a family she said they'd met on the desert going back for water and grass 'cause they'd missed their way. She said there was only one wagon, and there was a woman and child in it; and she knew they could never get through ... without help. We told her we had no orders to go any farther then. She said she didn't care for orders. She said she didn't believe anybody would blame us for doing what we were sent out to do. . . . You see I've got a wife and little girl of my own; so I felt just how it was.

The men explained the situation. The recent snowstorm had cleared, but another could come at any time and seal the pass for the season. The Royces must abandon the wagon. It would slow their progress to a crawl on the rough ascents ahead. Besides, their four weakened oxen would never manage the final, steepest pull near the summit pass. The Royces must leave the wagon and move on with all haste, packing a few essentials on the backs of the animals. That night of October 12, Sarah reflected:

> I lay down to sleep for the last time in the wagon that had proved such a shelter for months past. I remembered well, how dreary it had seemed, on the first night of our journey (which now seemed so long ago) to have

only a wagon for shelter. Now we were not going to have even that. But, never mind, if we might only reach in safety the other foot of the mountains, all these privations would in their turn look small.

The next day the Royces packed what they could on their four oxen and one old horse, as well as two mules that the government riders lent them. They moved swiftly now, and by October 17 they were approaching the final, roughest part of the ascent. The trail went up a narrow canyon boxed in by high walls and plugged with massive granite boulders. By the next evening, they had neared the mountain crest. They slept near snowbanks from the recent storm. Water froze in every container. But the skies stayed clear. The next morning, October 19, Sarah ascended the final heights.

> Whence I looked, down, far over constantly descending hills, to where a soft haze sent up a warm, rosy glow that seemed to me a smile of welcome; while beyond, occasional faint outlines of other mountains appeared; and I knew I was looking across the Sacramento Valley.
>
> California, land of sunny skies—that was my first look into your smiling face. I loved you from that moment, for you seemed to welcome me with loving look into rest and safety.

It took several days to make the descent. One week later heavy snows sealed the Sierra Nevada passes for the winter. The Royces were on the right side of the mountain. They started a new life in the hardscrabble mining towns springing up in the western Sierra Nevada foothills. In the years ahead, Mary would learn to read with a book called *Little Ella*.

THE FORTY-MILE DESERT and Sierra Nevada crossings were fearsome ordeals for nearly all California-bound emigrants. But these hardships were just part of a four-month, 2,000-mile journey.

It began with a 700-mile crossing of the Great Plains—easy stuff compared to what would follow. At the western edge of those plains, they entered the Rocky Mountains, the beginning of the North American Cordillera—the great mountain belt that stretches from the Rockies to the Pacific coast. To reach California, they had to pass through five Cordilleran geologic provinces: the Foreland Ranges of the Rockies, the Overthrust Belt, the Snake River Plain,[1] the Basin and Range/Great Basin, and the Sierra Nevada. Each province slung its own peculiar arrows of outrageous fortune at those passing through. Each one evoked wonder,

---

1. Emigrants who took the Hastings Cutoff, Salt Lake Cutoff, or Hudspeth Cutoff did not pass by the Snake River Plain.

joy, fear, or detestation, depending on circumstances. And each has a marvelous scientific story to tell.

THE EMIGRANTS set out once the warmth of spring had pushed winter off the Great Plains and the young grass needed for the livestock had sprung up. They headed upstream along the valleys of the Platte River and North Platte River across present Nebraska. The Great Plains lie on a stack of sedimentary layers, several thousand feet thick, shed east from the Rocky Mountains. The layers rise and thicken to the west, making a smoothly ascending ramp that took the emigrants steadily uphill to the foothills of the Rockies. Deep below the plains, the continental basement—the crystalline rock that makes up the foundation of North America—bears evidence of titanic collisions between small blocks of primordial crust that built the core of the continent nearly 2 billion years ago.

Moving west across the Great Plains, the emigrants saw aridity slowly wrap its tendrils around the land. There were fewer trees, and then none at all. Rolling grasslands stretched to the horizon, interrupted only by passing buffalo herds. In the valley of the North Platte River, they came into a landscape of stunning rock formations—stony vanguards of the great mountains that still lay ahead. First Courthouse Rock and Jail Rock loomed up, then Chimney Rock, Castle Rock, and Scotts Bluff. "No conception can be formed of the magnitude of this grand work of nature [Chimney Rock] until you stand at its base & look up," forty-niner Elisha Perkins marveled. "If a man does not feel like an insect then I don't know when he should."

Although massive on a human scale, the rock monuments of the North Platte Valley are but puny remnants of sedimentary layers that once stacked up so high on the Great Plains that they lapped at the chins of the highest Rocky Mountain peaks to the west. Several million years ago, the ancestral rivers of the plains began to eat into these layers, carving them away from the mountains. The rivers left a few scraps, standing today as isolated monuments high above the denuded landscape. We see the Rockies rising abruptly from the Great Plains today because these rivers have exhumed the mountains from deep burial.

The emigrants entered the Rockies in present southeastern Wyoming as they followed the North Platte River around the north end of the Laramie Range. Here they wrote with amazement of the tortured rocks—bent, broken, tilted up on edge—products of the grand geologic violence that spawned the Rockies. W. S. McBride, an 1850 emigrant, gazed at distorted rock layers, "standing edgewise . . . thrust through the earth's surface by some convulsion or subterranean force." These easternmost uplifts of the Rocky Mountains are called the Foreland Ranges. Each is made of a distinct block of basement rock squeezed up thousands of feet along faults by colossal sideways compression of the Earth's crust. Broad basins, 20 to 100 miles across, separate the ranges. As the deep basement blocks rose, they bowed up the sedimentary layers overhead so that today you see these layers leaning up against the mountains like boards stacked against the walls of mighty houses.

The Foreland Ranges face the Great Plains like a great wall. But a wide gap in the wall exists in Wyoming, between the Laramie, Bighorn, and Wind River ranges. This is why the Oregon-California Trail passed through here. By following the valleys of the North Platte River and then the Sweetwater River west through this gap, the emigrants could ascend gradual slopes all the way to the Continental Divide at South Pass. It was a good travel plan, as long as the rivers cooperated. But in some places, the rivers slash deep canyons straight through ridges and uplifts — even where a clear route around lies nearby. Faced with these impassable canyons, where the water thrashes against vertical walls, the emigrants had to detour. Where the North Platte River cuts through the Hartville Uplift, forty-niner William Swain endured a weeklong detour through "a broken, rocky, mountainous country [where the] road has been strewn with articles left by the emigrants to lighten their loads."

The oddity of rivers going *through* mountain ridges rather than *around* bothered many emigrants. Why would a river cut through millions of tons of solid rock to go through an obstacle when, in theory, it could have gone another way? Contemplating the Sweetwater River at Devils Gate, where the river punches straight through a granite ridge, forty-niner A. J. McCall wrote, "It is difficult to account for the river having forced its passage through rocks at this point when a few rods south is an open level plain over which the road passes." The puzzle is solved when we realize that the rivers once flowed high *above* the ridges that they cut through today. Thousands of feet of sand and gravel once covered all but the tallest Rocky Mountain ranges. Rivers wandered this ancient gravel plain, oblivious to ridges buried far below. A few million years ago, rejuvenated by uplift of the region or wetter climates, the rivers began to flow faster and bite downward into the sand and gravel layers. Where the downcutting rivers met long-buried ridges, they sliced on through to establish the paths that we see today.

Onward, westward, and upward. Leaving the North Platte River, the emigrants crossed overland to the Sweetwater River and followed its smoothly ascending valley upstream to the Continental Divide at South Pass, 7,550 feet above sea level, in present southwestern Wyoming. Here, at this "elevated and notable back-bone of Uncle Sam's," they celebrated. They were halfway to California — 1,000 more miles to go. Some of them may have seen South Pass as a divine validation of Manifest Destiny. The gentle ascent and the relatively low elevation seemed to mark the pass as God's natural gateway through the Rocky Mountains.

Whether by God's hand or nature's, South Pass exists by geologic consent. After the Foreland Ranges squeezed upward, a massive mountain blocked the way to South Pass. Later, as the crust stretched, this mountain slid down like a wedge between two large faults to form the Sweetwater Valley, opening the way west through South Pass. Had this not occurred, there would be an unbroken mountain barricade from Montana to New Mexico, and the Oregon-California Trail—and America's westward expansion—would not exist as we know it.

Crossing the Continental Divide, the emigrants drank for the first time from waters that flow west to the Pacific Ocean. Spirits were high. The first 1,000 miles were over, and it hadn't been that bad.

UP TO SOUTH PASS, the emigrants had ascended the valleys of east-flowing rivers—the Platte, the North Platte, and the Sweetwater. The rivers gave water, the grassy bottomlands provided feed for livestock, and the smooth slopes of the valleys made natural avenues for overland travel. West of South Pass, the land becomes less cooperative. The grain of the landscape—the trend of rivers, canyons, and ridges—tends to be north-south, cutting across westerly routes of travel. The reason comes down to North America's own history of westward migration. For the past 200 million years, our continent has pushed west, overriding several thousand miles of ancient seabed. The result is a landscape of north-south-trending mountains that stretches from Wyoming to California. The proliferation of trail cutoffs west of South Pass reflects this shift in the landscape. Up to South Pass, there was one road. Beyond the pass, the overland trails look like a rope pulled apart in the middle, with strands splitting off from the main trail and then rejoining it tens or hundreds of miles farther west. Some of these cutoffs shortened distances through the rough land. Others were worse than useless, saving no time at the cost of harder travel.

The emigrants descended from South Pass into the Green River Basin—6,000 square miles of desolate sagebrush wilderness flanked by mountains and riven by north-south-trending canyons. The Green River flows south across the basin and exits through the Uinta Mountains, joining the Colorado River on the other side. The Green and its many tributaries have sliced deeply into the flat sedimentary layers that fill the basin, cutting dozens of steep-walled valleys. Down into the valleys the emigrants skidded—and up they labored on the far sides. In between, they crossed miles of parched plains, teeth set against the wind. "It had been so windy and dusty today that we sometimes could scarcely see the length of the team, and it blows so tonight that we cannot set the tent or get any supper, so we take a cold bite and go to bed in the wagons. The wagons are anchored by driving stakes in the ground and fastening the wagon wheels to them with ox chains." That was how 17-year-old Eliza Ann McAuley celebrated Independence Day, July 4, 1852.

With relief, they passed from the Green River Basin into the Overthrust Belt, along the present Idaho-Wyoming border. Here they climbed into a landscape of serial valleys and ridges that run north and south for miles. Like the folds in a carpet shoved against a wall, the Earth's crust in the Overthrust Belt has been pushed from the west and bent into parallel north-south ridges. Slabs of rock thousands of feet thick have broken free along ramplike faults and slid east for as much as 100 miles, stacking up on one another like overlapping shingles. The Overthrust Belt is a geologic fold-and-thrust belt—a mountain belt formed by horizontal squeezing and sliding of the upper few thousand feet of the Earth's crust.

The pine forests and sparkling streams of the Overthrust Belt were an improvement over the desiccated Green River Basin. But getting through the north-south ridges was miserable hard work. "The word steep does not begin to convey an idea of the roads," 1852 pioneer Enoch Conyers wrote. "Several times I felt sure the wagon would tip over on the tongue yoke of cattle" during the precipitous descents. They skidded down the worst slopes in a controlled free fall, slowing the wagons by locking the wheels with chains and pulling back mightily on ropes.

West of the Overthrust Belt, the emigrant trail clips across the northwestern corner of the Basin and Range Province before arriving at the southern edge of the Snake River Plain in present southern Idaho. This vast volcanic region stretches more than 500 miles from northwestern Wyoming across southern Idaho into Nevada and Oregon. It is a bleak, black-rock landscape, paved with basalt flows and dotted with volcanic cones. The Snake River cuts a steep-walled canyon several hundred feet deep into the lava beds. "Of all countries for barrenness I have ever seen, it certainly exceeds any. I doubt if it can be equaled in any part of this continent," Major Osbourne Cross declared. The black lava blasted summer heat like an oven. It was bitter irony that the Snake—the largest river the emigrants had seen since the Missouri River nearly 1,300 trails miles back—could provide little relief from thirst and heat. As Ezra Meeker explained, "In some places we could see the water of the Snake River winding through the lava gorges, but we could not reach it, as the river ran in the inaccessible depths of the canyon." Below the lava beds of the Snake River Plain lies a set of yawning volcanic craters, some of them 50 miles across. They speak of volcanic Armageddon—a 16-million-year history of repeated, life-incinerating eruptions whose latest creation is Yellowstone National Park.

Where the Raft River joins the Snake River from the south, California-bound emigrants bid farewell to their Oregon-bound trail brethren. "'The Oregon Trail' strikes off to the right & leaves us alone in our glory, with no other goal before us but Death or the Diggins," forty-niner Wakeman Bryarly noted at this parting of the ways. The Oregonians continued west along the Snake, while the Californians turned south up the Raft River Valley, bound for the headwaters of the Humboldt River in present northeastern Nevada. They now faced a 600-mile push through the Basin and Range Province, most of it across the heart of present Nevada. Only the 700-mile crossing of the Great Plains represented a longer haul within a single geologic province. But travel across the Basin and Range was a far cry from the easygoing trek across the Great Plains. It would be hard to imagine worse country for east-west travel. The Basin and Range takes its name from its topography—dozens of long, narrow mountain ranges separated by arid, gravel-filled basins. Basin, range, basin, range, lined up north-south from western Utah clear across Nevada to the Sierra Nevada. From high and distant vistas, the emigrants stared glumly west at range after range, cresting toothy and raw to the horizon like rock waves on a frozen sea. John Hawkins Clark gave this report while crossing the Tuscarora Mountains in 1852:

> Our road this afternoon is up a steep mountain side seven miles long; the steepest, roughest, most desolate road that can be imagined. The mountains that border this valley . . . have a decrepit and worn-out look. . . . It makes a man lonesome and homesick to contemplate their forlorn, deserted and uncanny appearance. Stunted and scattered cedar trees, broken down by the snows and wild winds of the winter season, gives them a sort of ghost-like appearance that makes one shudder to behold.

Most mountain ranges, including the Foreland Ranges and the Over-thrust Belt of the Rockies, form through horizontal compression of the Earth's crust—sideways squeezing that thickens the crust and pushes up mountains. The Basin and Range is different. Here the Earth's crust has stretched more than 200 miles east to west, breaking up into north-south-trending basins and ranges. For the emigrants, there was only one viable path west through this gauntlet of ranges—the Humboldt River. The Humboldt snakes west for 350 miles across Nevada, nosing its way around the ends of the north-south ranges or cutting through them. It is the only permanent river crossing the Great Basin—a vast region of internal drainage *within* the Basin and Range Province where rivers have no outlet to the sea. The Humboldt was the emigrants' lifeline across the

Great Basin, the only route with reliable water and grass. But what a river—muddy and sullen, foul-tasting and salty from evaporation, shrinking downstream before expiring at the Humboldt Sink, 40 miles shy of the Sierra Nevada. And the surrounding country was no better. "Nothing but the hot sterile lands and dust immediately around us & naught in the distance to relieve the eye, but bare rugged hills of basalt," forty-niner Bennett Clark groaned. "Our feelings now [are] that if we once get safely out of this great Basin we will not be caught here again in a hurry."

Once the Humboldt died, the going only got worse. From the Humboldt Sink, the emigrants had to make a near-waterless leap—a leap of faith—across the hottest, driest section of the trail: the Forty-Mile Desert. It was 40 miles from the end of the Humboldt River to either the Truckee River or the Carson River—the nearest streams flowing off the eastern slopes of the Sierra Nevada. Here the grim trails were easy to find, marked by bloated carcasses and bleaching bones of hundreds of livestock that had perished from heat and thirst. Their decomposing bodies filled the air with the stench of death. Desperate emigrants discarded tons of valuable possessions to lighten loads for weakened animal teams. Wagons and goods piled up, abandoned, in the desert. The only water, barely potable, lay in a handful of dug wells clogged with algae, or in sulfurous holes that belched steam like portals to hell. Aridity reigns in the Forty-Mile Desert—and across the rest of the Great Basin—because of the Sierra Nevada. The mountain traps cloud moisture on its western flanks. To the east, downwind of its rain shadow, a desert now stretches for hundreds of miles across five western states.

The Sierra Nevada greeted those who stumbled, with cracked lips and swollen tongues, out of the Forty-Mile Desert. Here loomed the final and greatest barrier of the journey, with passes as high as 9,500 feet that threatened snow and a frozen fate to stragglers. This great block of the Earth's crust began to rear up in earnest about 5 million years ago, and it continues to rise today. Earthquakes pop off periodically along the faults that bound the range. Each one lifts the mountain a bit higher. The quakes are part of the ongoing stretching of the Basin and Range Province—a process that each year pulls Sacramento away from Salt Lake City by nearly a half-inch. On the other side of the Sierra Nevada, scattered through its western foothills, lay the gold that all hoped would "repay them for all their hardship and suffering"—gold that migrated to California long ago through the quirks of geologic history.

THE EMIGRANTS could not know of these links between the history of the North American continent and their experiences on the road west. Geology was an infant science in emigrant days. It sprouted from the seeds of deep time germinating in the minds of European thinkers during the seventeenth and eighteenth centuries. It grew up during the nineteenth and early twentieth centuries with exploration, mapping, cataloging, and observing, and burst into maturity in recent decades with a unifying theory called plate tectonics.

Nonetheless, many emigrants were intensely curious about the geologic landscape. They wanted to understand the rocks, rivers, and mountains. The landscape around them changed as they went west, but its influence on their lives was constant. The land occupied their thoughts during the day and often dominated their writings at night. Sometimes the landscape inspired rapture, at other times loathing. In June 1850 by Chimney Rock, emigrant Dan Carpenter wrote, "From the top of the bluff near the chimney I had a splendid view westward of some of the most beautiful wild and romantic scenery as I ever beheld." Some two months later, along the Humboldt River, he was less enchanted. "This is the poorest and most worthless country that man ever saw—No man that ever saw has any idea what kind of a barren, worthless, valueless, d—d mean God forsaken country there is, . . . not God forsaken for He never had anything to do with it." Carpenter's sentiments mirror the geologic diversity of the West—a land of warts and wonder, simultaneously hostile and sublime, spectacular and severe.

BETWEEN WINTER'S
CHILL BRACKETS

The gold mania rages with intense vigor, and is carrying off its victims hourly and daily. . . . [O]ur young men—including mechanics,
doctors, lawyers, and we may add, clergymen—are taking leave of
old associations, and embarking for the land of wealth, where the
only capital required to make a fortune is a spade, a sieve, a tin
colander, and a small stock of patience and industry.

NEW YORK TRIBUNE, December 11, 1848

The overland journey to California followed a tight schedule bracketed
by winter. Emigrants could not set out across the Great Plains until late
April or early May. Trails turned to mud by the spring thaw needed to dry
out, and the spring grass had to come up, for grass was fuel in this age of
animal power. On the far end of the journey, the Sierra Nevada had to be
crossed before the first snows threatened to seal the high passes, which
could happen as early as October. That left a five-month window—May
through September—to complete the journey, perhaps six months if the
weather on either end cooperated. The constraints of winter created a
singular disadvantage in timing. Some of the hardest sections of the journey—the deserts of the Great Basin—had to be crossed at the absolutely
hottest time of year, in August and September.

Wagon travel averaged about 15 miles per day. Good travel days saw
more than 20 miles go by, but river crossings, bad weather, or rough terrain slowed progress on some days to a handful of miles. Some days were
spent resting or making repairs. Many emigrants halted on Sundays to
observe the Sabbath. All in all, the 2,000-mile trip typically took from
four to five months by ox-drawn wagon, three to four months by pack
train.

A successful overland crossing took more than timing. It took preparation and planning. In the early 1840s, at the start of major overland
emigration to Oregon and California, the United States ended at the Missouri River. When emigrants set foot on the Missouri's west bank, they
entered wilderness—at least from the perspective of white Americans.

While they had many opportunities to trade with Indians along the way, they could not count on getting essentials like gunpowder, flour, or fresh animals anywhere between the Missouri River and trail's end. The only sizable town along the way was Salt Lake City, established by the Mormons in 1847, and only emigrants who took the Salt Lake Cutoff or Hastings Cutoff passed that way. A handful of trading posts—Fort Laramie, Fort Bridger, and Fort Hall, in particular—lay scattered along the route, and goods there were highly priced and often in short supply.

Vast portions of the West remained unexplored and unmapped in the mid-nineteenth century. Existing maps were often unreliable. "I do not know where we are, nor do our maps," one forty-niner exclaimed along the Hudspeth Cutoff. "We have concluded to throw away the maps, trust in good luck and when we arrive in California we shall probably know it." In the decades before the great emigration, mountain men—trapping beaver and trading with resident Indians—had blazed a handful of overland routes, but these trails were often impassable for wagons. John C. Frémont's government-sponsored exploring expeditions of 1842 to 1845 coincided with the early years of emigration to Oregon and California. The earliest emigrants were ahead of Frémont. (Frémont, a savvy self-promoter, gained national stature and the appellation "Pathfinder" by publicizing his expeditions. The Pathfinder actually found few paths that someone else had not found first, although he did produce valuable maps and guides for later travelers of those routes.) For much of the trip, an emigrant's only knowledge of what lay ahead came from Frémont's published works or brief descriptions in thin trail guidebooks.

In addition to timing, preparation, planning, and pluck, a successful overland journey took some luck, and many people did not bring enough. Roughly 20,000 died on the overland trails to California between 1840 and 1859—an average of 10 graves per mile.[1] They died mostly of disease, particularly cholera from contaminated water along the Platte and North Platte rivers. Poor sanitation and the burial of infected bodies near watercourses helped spread the disease. Cholera worked quickly, killing by dehydration. Accounts tell of people waking at dawn feeling fine and lying dead by sundown. They also died of dysentery, tuberculosis, smallpox, mumps, pneumonia, and "mountain fever" (probably a tick-related disease). About twice as many emigrants fell from disease as died from all other causes of death combined. But even if you escaped disease, there were plenty of other ways to die on the road west, including ac-

---

1. It is difficult to get precise estimates of the number of emigrant trail deaths. Historian John Unruh estimates 15,000 to 30,000 (Unruh 1979, 408, 516n75), so 20,000 is probably a reasonable figure.

2.1 A pioneer family crossing the Great Plains. Courtesy of the Denver Public Library, Western History Collection, X-11929.

cidental gunshot, trampling or kicking by animals, and being crushed under wagon wheels. Many drowned on river crossings. The Platte, North Platte, and Green rivers claimed the most victims, but some even drowned on smaller rivers like the Humboldt. Some were killed by Indians. Others were shot, stabbed, or bludgeoned to death in fights with other emigrants.

Not all was hardship on the overland trails. There was plenty of fun: music, dancing, hunting, socializing around campfires, and sex too— probably more than mid-nineteenth-century sensibilities would admit in their journals and memoirs. As John Lewis reported from the Great Plains, "Love is hotter her[e] than anywhere that I have seen when they love here they love with all thare mite & some times a little harder." Moreover, the spectacular western landscape repaid some emigrants for the journey's hardship and drudgery. "The scenery through which we are constantly passing is so wild and magnificently grand that it elevates the soul from earth to heaven and causes such an elasticity of mind that I forget I am old," Harriet Ward wrote in 1853. (She was fifty.) The unfolding land ahead was so big, so wild, so different from the domesticated land left behind—shimmering prairie grasslands, sawtoothed mountains swathed in perpetual snows, bizarre sculptural rock formations. For 1850 emigrant Byron McKinstry, "The remembrance of scenery, at times, beautiful, picturesque, sublime, is all the compensation I receive for present toil. I could sit on some mountainside for hours and gaze with rapture on the new scene now spread before me—the whole forming a picture more beautiful than ever emanated from the hand of painter, though made up of rocky mountains and barren plains." Upon first seeing the Rocky Mountains, forty-niner A. J. McCall declared simply, "I do not know when I have witnessed a more beautiful sight."

Hardship floats up though, eventually, in all emigrant accounts of the journey west. "To enjoy such a trip," one pioneer wrote, ". . . a man must be able to endure heat like a Salamander, mud and water like a muskrat, dust like a toad, and labor like a jackass. He must learn to eat with his unwashed fingers, drink out of the same vessel with his mules, sleep on the ground when it rains, and share his blanket with vermin. . . . He must cease to think, except as to where he may find grass and water and a good camping place. It is hardship without glory, to be sick without a home, to die and be buried like a dog." Women, many of whom were uprooted from homes and communities by husbands with dreams of land or gold, often felt the hardship with special poignancy.[2] Here, for instance, is 1860 emigrant Lavinia Porter on the high plains of present Wyoming.

> I would make a brave effort to be cheerful and patient until the camp work was done. Then starting out ahead of the team and my men folks, when I thought I had gone beyond hearing distance, I would throw myself down on the unfriendly desert and give way like a child to sobs and tears, wishing myself back home with my friends and chiding myself for consenting to take this wild goose chase.

Emigrants had a saying for intense hardship. They called it "seeing the elephant." Almost everyone saw the elephant somewhere along the overland trail—perhaps in the Black Hills (the foothills of the Laramie Range), or on the Sublette Cutoff, or the rough ridges of the Overthrust Belt. If not

2. The historian John Faragher, based on comparative research of men's and women's accounts of the westward journey, has concluded that "not one wife initiated the idea [of emigrating to Oregon or California]; it was always the husband. Less than a quarter of the women writers recorded agreeing with their restless husbands; most of them accepted it as a husband-made decision to which they could only acquiesce. But nearly a third wrote of their objections and how they moved only reluctantly" (Faragher 1979, 163).

2.2 An encounter with the elephant—the quintessential symbol of hardship on the overland trail. William B. McMurtrie, 1850, E. F. Butler, San Francisco Fine Arts Collection, Lettersheet B-65. Courtesy of California Historical Society, FN-04479.

there, then certainly along the Humboldt River, the Forty-Mile Desert, or the appalling Sierra Nevada passes. "Oh, surely we are seeing the elephant, from the tip of his trunk to the end of his tail!" Lucy Cooke wrote at a fearful river crossing. After a Great Plains hailstorm had lacerated shoulders and backs to bleeding, James Lyon decided that the "storm was decidedly severe, a touch of the terrific, something of the elephant." Wandering past heaps of dead cattle and abandoned provisions in the Forty-Mile Desert, Lucius Fairchild concluded, "That desert is truly the great Elephant of the route and God knows I never want to see it again." The elephant chased some people home. John Edwin Banks saw a rather rare sight in 1849—a man heading east. "[He] says he can't go all the way. Has enough money; loves his wife more than gold."

2.3 A yoke of oxen is two animals leashed together by a yoke: a crossbar of carved wood fastened to their necks with oxbows. Two or three yoke (four to six animals) pulled a typical emigrant wagon. Most emigrants brought along several additional animals—the nineteenth-century equivalent of spare tires. Of oxen, mules, and horses (the three animal engines of the westward migration), oxen were by far the most common. Although slow, oxen were relatively inexpensive, immensely strong, less likely than horses or mules to be stolen by Indians, and could subsist reasonably well on available grass. (Visitors Center at Scotts Bluff National Monument, western Nebraska.)

MOST EMIGRANTS walked much of the 2,000 miles from the Missouri River to California. Regular riding in the wagons was only for the very young, pregnant, sick, or injured. Most emigrant wagons had no springs, and riding was often uncomfortably rough. Walking avoided the continuous clouds of dust kicked up by wheels and hooves. Most important, the emigrants walked because they knew that the fortunes of their journey depended on the health of their animals. Livestock already had a heavy task hauling all the supplies for the journey plus the basic possessions for beginning a new life on the other end. "The fact is every attention to your cattle is actualy necessary to take you through this trip," Richard May explained in 1848, adding, "Oxen are the Centeral object on this route, you belong to them instead of them to you." Forty-niner James Pritchard echoed these sentiments. "It has been and should have been our object from the start to preserve the strength of our teams as much as the nature of the case would allow. It is an alarming and fearful thing to see as we do every day teams broken down by the mismanagement of their owners in this remote wilderness."

Alarming and fearful, indeed, for the loss of a livestock team transformed the journey from one aimed at starting a new life to one of base survival. You need only think about how far you could walk with a maxi-

2.4 The wagon in the foreground is typical of the type used by emigrants. It was light-weight and small, with a bed measuring about four feet wide and ten feet long, and designed to be pulled by two or three yoke of oxen or, less commonly, three to five yoke of mules. The larger, heavier wagon in the background is a Conestoga, a freight wagon requiring eight or ten animals, with a curving bottom to keep casks and other large freight items from shifting. Conestoga wagons were too large and heavy for most emigrants. (Display outside the Visitors Center at Scotts Bluff National Monument, western Nebraska.)

mum load of food and clothing on your back to realize how quickly the claws of starvation reached for those emigrants whose animals died. To lighten wagons for weakened or reduced livestock teams, emigrants discarded heaps of possessions, particularly in the final stretches of the journey. Anvils, kegs, dressers, grinding stones, trunks, plows, drills, augers, baking ovens, harnesses, chains, mattresses, quilts, saws, planes, scythes, crockery, mining tools—valuable items hauled all the way from the Missouri River—were tossed out in the wilderness. Opportunists scavenged many of these discarded goods, which often commanded top prices in Salt Lake City or in California, and souvenir hunters in more recent years have (illegally) scooped up many others. But even today along the traces of the trails, you can still find bits of crockery, square nails, fragments of iron wagon wheel tires, rusted sections of barrel hoops, and other bits and pieces. And you can never know—is this the simple detritus of a successful journey, or was it thrown overboard during a worrisome and desperate struggle?

THE OREGON and California trails form a 2,000-mile-long letter Y tipped west. The base of the Y, the beginning of both trails, is represented by several jumping-off points on the Missouri River in present Kansas and Nebraska. The fork of the Y lies 1,300 miles west, on the Snake River in present southern Idaho. Seven hundred miles along the right fork takes you to Oregon's Willamette Valley. The left fork, also 700 miles, takes you to the gold fields of the western Sierra Nevada foothills. Legend has it that at the fork there was a sign pointing right, "To Oregon," for those who could read. The left fork was signed only by a pile of rock rubble—illiterate gold-grubbers go this way.[3] Because of the gold rush, more emigrants ultimately traveled to California than to Oregon on the overland trail system. But prior to 1849 most emigrants were Oregon-bound, and so the route west became generally known as the Oregon Trail. Oregon-California Trail is the designation favored by most historians and trail buffs today. Still others prefer Oregon-California-Mormon Pioneer Trail, to acknowledge the emigration of the Latter-day Saints to the Salt Lake Valley beginning in 1847.

Whether bound for Oregon, California, or the Salt Lake Valley, everyone took more or less the same route for the first half of the journey (see the "Missouri River to South Pass" map on pp. viii–ix). They approached the Platte River from six major jumping-off points on the Missouri River, and then followed the valleys of the Platte and the North Platte rivers west across the Great Plains. They continued along the North Platte around the Laramie Range (the beginning of the Rocky Mountains), crossed over to the valley of the Sweetwater River, and then followed the Sweetwater up to South Pass, in present southwestern Wyoming. South Pass marked the halfway point of the journey—about 1,000 miles behind, 1,000 more to go.

Beyond South Pass, the trails split, offering several choices about which way to go (see the "South Pass to Sierra Nevada" map on pp. x–xi). California-bound emigrants all had the same goal—to get to the Humboldt River in present northern Nevada. From South Pass, everyone bound for California had to decide which route to take to get to the Humboldt. It was about 500 miles, give or take 50, whichever way you went. There was no good way.

The oldest route, and the most indirect, followed the original Oregon

3. Some accounts place this sign at the fork of the Sublette Cutoff, an important trail fork about 20 miles west of South Pass.

Trail for much of the way. It headed southwest from South Pass across the Green River Basin to Fort Bridger, a trading post opened in 1841 by mountain man Jim Bridger. The trail then cut northwest across the Overthrust Belt and followed the Bear River north to Soda Springs (in present southeastern Idaho). It continued into the volcanic country around the Snake River, and then headed west along the Snake to the Raft River. There it split from the trail to Oregon and turned south through the Basin and Range Province to the headwaters of the Humboldt River. Going this way, it was about 550 miles from South Pass to the Humboldt—nearly twice as far as a straight line.

Because this original California Trail was so indirect, explorers during the 1840s blazed several alternative routes, or "cutoffs," at various points between South Pass and the Humboldt River. By 1849 most emigrants were taking one or more of these cutoffs instead of the original trail. The main cutoffs, with the year of first use, are the Sublette (1844), Hastings (1846), Salt Lake (1848), and Hudspeth (1849). The Sublette Cutoff was the only one that saved significant time and distance. From near South Pass, it headed straight west across the Green River Basin and the Overthrust Belt, connecting up to the original trail again at the Bear River in present westernmost Wyoming. By cutting out the dogleg down to Fort Bridger, the Sublette Cutoff shaved off about 50 miles and three or four days of travel. The way was rough, and it had a grim run of 45 miles without water. Many cattle died on this stretch, leading to the joke about how easy the Sublette Cutoff was to follow—just look for the path of bleached bones.

The Hudspeth Cutoff looked good on a map. Like the Sublette Cutoff, it cut out a dogleg, bypassing the looping route of the original trail north to the Snake River. The cutoff headed straight west from Soda Springs, cutting across the Basin and Range ridges to the Raft River, where it rejoined the original California Trail near a natural landmark called the City of Rocks. In reality, the Hudspeth Cutoff was no cutoff at all. The slower travel up and over the multiple mountain ridges burned up much of the time saved by the minimally shorter distance. The uselessness of the cutoff hit home for emigrants who waved good-bye to companions opting for the longer route up to the Snake River, only to meet them again where the trails rejoined some 10 days later. Nonetheless, the Hudspeth Cutoff saw heavy use from 1849 on.

To take the Salt Lake Cutoff, emigrants followed the route of the Mormon pioneers southwest from Fort Bridger down to the present site of Salt Lake City. They then turned northwest around the east and north sides of Great Salt Lake, joining the original California Trail near the City of Rocks in present southern Idaho. The Salt Lake Cutoff was not appreciably shorter, but it had the advantage of passing through Salt Lake City, where one could buy supplies and trade trail-worn livestock for fresh ones. (The Mormons at Salt Lake City had a smart business operation, buying worn-down emigrant livestock for cheap, freshening them up in pasture, and then selling them at a profit to later emigrants.) Some emigrants even opted to over-winter with the Mormons, continuing on to California the next spring. Most passed on through, though, either because they didn't care for Mormons or because they were in a rush for gold.

The Hastings Cutoff, like the Salt Lake Cutoff, began at Salt Lake City. It went around the south end of Great Salt Lake and headed west across the Salt Lake Desert, eventually meeting the Humboldt River a few miles downstream of present Elko, Nevada. It was a terrible route. Although no one realized it at the time, it was just as long as the original trail, perhaps a bit longer—*and* it was much harder, involving a ghastly run of 80 waterless miles across the Salt Lake Desert (in the area of today's Bonneville Salt Flats). In 1846 promoter Lansford Hastings convinced several parties of late-starting emigrants to try his new cutoff, touting it as a faster way to California. The Donner party was on the tail end of this group. The parties ahead of the Donners made it over the Sierra Nevada before winter snows. But for the Donner party, the Hastings Cutoff was a disaster, slowing them down, killing their livestock, and exhausting minds and bodies alike. The delayed party began their ascent of the Sierra Nevada late in October 1846. Early snows trapped them, and their cannibalistic fate horrified the nation. The Donner debacle rightly sullied the reputation of the Hastings Cutoff, and the route saw only moderate use in later years. Monday-morning quarterbacks of history will always debate whether the delays of the Hastings Cutoff doomed the Donners. Maybe they would have been slow no matter which route they took. One thing is certain—there was no good way to get to California overland in the mid-nineteenth century. West of South Pass, it was hard no matter which way you went.

Once on the Humboldt River, emigrants had two options. They could follow the Humboldt for all of its desolate 350 miles to the Humboldt Sink, which dumped them out at the Forty-Mile Desert. They could then cross the desert and ascend the Sierra Nevada over one of several passes. Going this way meant making a frontal assault on the Sierra Nevada's high eastern escarpment. The main trails for this were the Truckee Route over Donner Pass and the Carson Route over Carson Pass. The alternative to the frontal assault was to make a long end run around the northern end of the range. For this option, emigrants peeled away from the Humboldt River about 90 miles above the Humboldt Sink and headed northwest on the Applegate-Lassen Trail. This option added as much as one month to the journey, but it avoided the high passes to the south.

*

The roughness and aridity of the West poured out a world of rock to emigrant eyes. Rock—naked, broken, mountainous rock—*is* the West. Water is sparse in much of the West. Soils are thin, plants few. Rock rises to view with rare clarity, revealing its intricacy and structure. For the emigrants, the dramatic exposures—in mountainsides, canyon walls, river valleys, and volcanic plains—stirred wonder and speculation. "This whole region of the world has been an enormous furnace and its power and force incomprehensible," forty-niner Joseph Middleton wrote of the Rockies. What forces, he wondered, could "roast such an immense tract of country" with vast outpourings of volcanic lava, and raise "horizontal rocks into almost perpendicular ridges that run north and south for I don't know how many miles and from far east on the headwaters of the Missouri to the Pacific Ocean."

The emigrants' journey unfolded within the constraints of time—the seasonal window between the spring thaw of the Great Plains and the autumn snows of the Sierra Nevada. Geology blossomed as a science when it shrugged off the constraints of time. Nearly everything important in geology traces back to the recognition of *deep time*—the realization that the Earth is very old, and that its long history can be read directly from rocks and fossils without reference to the Bible. As the late geologist Stephen Jay Gould has written, "All geologists know in their bones that nothing else from our profession has ever mattered so much."

By the time of the California gold rush, this revolution in thought about geologic time and the age of the Earth was in full swing in Europe. One of the most important books in the history of geology—Charles Lyell's *Principles of Geology*—first appeared in 1830, and at least one California-bound emigrant (John Edwin Banks) owned a copy. Lyell stood on the shoulders of a fellow Scot, James Hutton, and they both in turn erected the new science on a foundation laid more than a century earlier by a Dane named Nicolaus Steno.

Although an anatomist by profession, Nicolaus Steno was intensely curious about all things in the natural world, particularly rocks. In his 1678 book *De solido*, Steno proposed fundamental rules for how strata (layers of sedimentary rock like sandstone, limestone, and shale) form. Steno suggested that rock layers accumulate, one on top of another, in horizontal layers that extend laterally for long distances. Therefore, he reasoned, if you see strata that are today not horizontal, but tilted up on edge, the layers must have been tilted after they formed. If you see strata ending abruptly at cliffs or valley walls, erosion must have cut them off after they formed. If strata high up in mountains contain fossil seashells, then the seas must have once been much higher or the land much lower. Steno's principles seem simple today, but they were revolutionary in his time because they meant that one could interpret the Earth's history directly from rocks, without the authority or assistance of the Bible.

This was a key break with intellectual tradition. Before Steno, scriptural chronology was the only acceptable way to interpret the Earth's history. In 1655 Irish archbishop James Ussher published a painstaking accounting of the age of the Earth deduced from biblical generations. Ussher's conclusion: God had created the Earth on October 23, 4004 B.C.[4] Whether or not they agreed with Ussher's precision, many seventeenth-century minds assumed that the Earth was but a few thousand years old, that the biblical flood was a real event, and that today's landscapes were largely a result of the flood. But after Steno, many thinkers found it difficult to shoehorn their observations into biblical chronology. The problem was *time*. Steno's principles implied time in amounts far beyond the range of biblical accounting. A lot of time seemed to be required for strata to be laid down, layer by layer, thousands of feet thick, under the oceans; time for the seas to withdraw and leave the strata exposed on dry land; time for the layers to be cut up by eroding rivers; time to push and tilt the strata into new orientations in mountains. Archbishop Ussher couldn't find enough time in the Bible to do these things.

Steno planted the seeds from which a new science would sprout — a science of the Earth divorced from scriptural time. During the eighteenth century, more and more scientists began to drift away from scriptural chronology as an explanation for the Earth's history. Nonetheless, it took more than a century after Steno for modern geology to blossom. The responsibility for that lay largely with James Hutton, his champion John Playfair, and his intellectual heir Charles Lyell. Nicolaus Steno cracked the door to deep time. Over the next century, various thinkers opened it wider. In 1785 James Hutton kicked it open permanently.

Hutton began his career as a medical doctor, but grew wealthy as a gentleman farmer and by his forties had the means to devote himself full-time to his greatest passion — the nascent science of geology. Farming in his native Scotland, Hutton had seen how erosion by muddy streams continuously wore down the land surface. He reasoned that if this process went on long enough, all land everywhere would be cut down to a flat

---

4. Geologists have parties that begin on October 22; at midnight they toast the anniversary of the Earth's formation.

plain near sea level. Yet all around him rose craggy peaks, many of which contained fossil seashells. Hutton developed a theory in which land and sea regularly changed places in an endless cycle of destruction and renewal. Land formed where ancient sea bottoms rose up, lifting seashells toward the clouds. The clouds rained down, wearing away the land. Rivers washed the eroded particles into the sea, where they compacted into layers of sedimentary rock on the seafloor. That seafloor rose in turn to become new land. "There is presently laying at the bottom of the ocean the foundation of future land," Hutton decided. What caused seafloors to rise and become land? Hutton thought that it might be the buoyant effects of the Earth's internal heat. "The strata formed at the bottom of the sea [have] been elevated [into land], as well as consolidated, by means of subterraneous heat." How long did this take? Hutton wasn't sure—but he was convinced it was far longer than anyone had previously supposed. "Time, which measures everything in our idea . . . is to nature endless and as nothing." He presented this theory in 1785 in a paper read before the Royal Society of Edinburgh.

Hutton's theory was seminal because it didn't invoke any supernatural or unobservable processes to explain the Earth's history. Even though Hutton believed that the Earth was created for a purpose (to support humankind), he believed that its operation was completely naturalistic. Moreover, he believed that the same natural processes of erosion, sedimentation, and uplift that we can see in action today are enough to explain the formation of continents, soil, valleys, mountains, and rocks—in short, just about everything about the Earth's past that needs explaining. All that is needed for these immensely slow processes to do their work is *time*. So much time that Hutton saw an Earth with "no vestige of a beginning,—no prospect of an end."

Perhaps, for Hutton, the Earth spoke most clearly of its deep past at Siccar Point, a wave-battered headland on Scotland's east coast. Here, in 1788, Hutton and his friend John Playfair (a professor of mathematics at the University of Edinburgh) found metamorphosed sandstone layers standing on end like a vertical deck of cards. On top of the beveled-off edges of these layers lay another stack of sandstone layers, gently inclined. The lower layers, now vertical, had originally formed as horizontal layers of muddy sand on the floor of an ancient sea. They had turned to stone under the weight of overlying layers. Pressure and uplifting movements had then slowly turned the horizontal layers to vertical as it squeezed them up into mountains. Erosion had then beveled these mountains down to irregular nubs. After that, the region had once again sunk below the sea, and new sandstone layers had piled up on the eroded edges of the older ones. These layers had then, in turn, been buried and turned to stone. Finally, the entire package had been lifted once again above the sea, where erosion exposed the whole affair to view. All in all, the exposure at Siccar Point testified to a stunning span of time. "The mind seemed to grow giddy by looking back so far into the abyss of time," John Playfair would later write of Siccar Point. Archbishop Ussher's scriptural chronology gave the Earth only 5,792 years (4004 B.C. to 1788 A.D.) to do this work. No, said Hutton, that was not enough time—not nearly.[5]

By all accounts, Hutton was a delightful person. Streams of excited ideas poured forth without letup from his balding head. With twinkling eye and boyish enthusiasm, he charmed nearly everyone he met. But he had little talent with the pen. His published works "produce a degree of obscurity astonishing to those who knew him," John Playfair admitted— and he was Hutton's close friend and greatest admirer. In 1795 Hutton published *Theory of the Earth, with Proofs and Illustrations*—a book that presented his theory of the great age of the Earth and its cyclic exchange of land and sea. It is, to put it mildly, a difficult read. Spilling into two volumes totaling some 1,200 pages, containing voluminous untranslated passages in French, and beleaguered with some of the most turgid English prose ever set to print, *Theory of the Earth* is doubtless one of the least-read important books in the history of science. Luckily for Hutton—and the future of geology—John Playfair was not only dazzled by Hutton's ideas but could explain them in plain English. In 1802, five years after Hutton's death, Playfair published *Illustrations of the Huttonian Theory of the Earth*, which translated Hutton's murk for everyone else. Thanks to Playfair, Hutton's principles gained a beachhead in the collective mind of science instead of drifting on the waves of intellectual obscurity.

To watch geologic thinking evolve through the early 1800s is to witness the slow, lurching triumph of Hutton—the growing acceptance that the Earth is very old and that slow, gradual geologic processes, doing their work over eons, can explain the major features of the planet. It took time (appropriately) for Hutton's ideas to win converts. Although biblical views of Earth history were on the wane well before Hutton, many at the turn of the nineteenth century still believed that Noah's flood had carved the Earth's surface and deposited layered sedimentary rocks. The notion that a single, globe-girdling ocean had created most of the Earth's geology

5. Notice that Hutton presented his theory to the Royal Society of Edinburgh in 1785, three years *before* he and Playfair found the exposure at Siccar Point. Some textbook accounts perpetuate the myth that the Siccar Point outcrop was Hutton's "Eureka!" moment, when he first realized that the Earth was incredibly old. In fact, Hutton had already deduced his theory—and the great age of the Earth that it implied—years before. The Siccar Point outcrop merely affirmed to him the basic correctness of his theory.

reached its pinnacle in the theory of Neptunism. Promoted with tireless zeal by the charismatic German mineralogist Abraham Gottlob Werner, Neptunism held that all rock on Earth had precipitated from an ocean that once covered everything. (Neptunism did not require that this global ocean be the same one upon which Noah floated—but the easy connection made biblical literalists happy.) The Neptunists loathed Hutton and his growing legions, who held that much rock is a product of heat, melting, and volcanism.

Attacks on Hutton came from another quarter as well, from a diverse group of thinkers sometimes called catastrophists. Most of them accepted Hutton's evidence that the Earth was very old. But, they said, even if the Earth is old, it does not necessarily follow that geologic processes occur at slow, steady, uniform rates. For instance, followers of Hutton (who came to be called gradualists) said that mountains rise gradually from the buoyant effects of the Earth's internal heat and the cumulative work of many earthquakes. But catastrophists held that mountains could appear in a virtual eye blink, carved out by titanic (although not biblical) floods or hoisted quickly during cataclysmic earthquakes.[6] Catastrophists reasoned that such rapid and vigorous geologic spasms would have been common in the Earth's youth, when the planet was hotter. Gradualists said such notions were speculative and untestable, and therefore unscientific. Catastrophists accused gradualists of unimaginative and hidebound thinking. Vitriolic debates ensued. Normally respectful men hurled venomous insults. Normally somber men grew incandescent with rage. Arguments and counterarguments surged back and forth about the origin of mountains, continents, ocean basins, and all forms of rock. Time was on the side of Hutton, though, largely because of Charles Lyell.

More than anyone, Charles Lyell is responsible for the widespread acceptance of Hutton's principles. Playfair established Hutton's beachhead, but Lyell led the subsequent advance. Lyell graduated from Oxford in 1816 and after a few unenthusiastic years practicing law, turned full-time to a geology career. In 1830 he produced the first volume of his now-famous *Principles of Geology*. The book grew rapidly to a multivolume work that went through no less than 12 editions over the next four decades. *Principles of Geology* was seminal for two main reasons. First, it was the first all-encompassing book of geology based on Hutton's principles, gathering under one roof virtually everything that was then known about the sci-

---

6. Many catastrophists invoked massive flooding to explain the Earth's geology. This has led to the misconception that scientific catastrophists were biblical literalists. In fact, most were not. The nineteenth-century scientific debate between gradualists and catastrophists was about how fast geologic processes happen, not the veracity of the Bible.

2.5 James Hutton (*left*), a Scottish medical doctor, is widely regarded as the father of modern geology (Museum Property, U.S. Geological Survey). Hutton argued from geologic evidence that the Earth is incalculably old. He proposed that the emission of heat from the Earth's interior causes uplift of the land, while the forces of erosion break it down, producing an everlasting cycle of change with "no vestige of a beginning—no prospect of an end." The bachelor Hutton lived with his three sisters until his death in 1797. The Scottish geologist Charles Lyell (*right*), author of the seminal *Principles of Geology* (1830–33), proposed that the forces molding the planet today have operated continually throughout its history, and therefore that observation of geologic processes in action at present holds the key understanding the Earth's past—a concept he called uniformity and today generally known as uniformitarianism. Lyell was knighted in 1848 and made a baronet in 1864. He died in 1875 and is buried in Westminster Abbey.

ence. Second, it gave geology an empirical method. Study what the Earth does today, Lyell preached, and you can understand its past because the Earth has obeyed the same physical rules and laws for all time—a principle we today call uniformitarianism. Lyell's subtitle captures this theme: *An Attempt to Explain the Former Changes of the Earth's Surface by Reference to Causes Now in Operation.* Here, for instance, is Lyell on mountains: "We know that one earthquake may raise the coast of Chile for a hundred miles to the average height of about five feet. A repetition of two thousand shocks of equal violence might produce a mountain chain [today's Andes] one hundred miles long, and ten thousand feet high." This is vintage Lyell. Look at what the Earth is doing today, extrapolate those processes back through the abyss of time, and—*voilà!*—mountains are born.

It would be hard to overstate the influence that *Principles of Geology* had on nineteenth-century science. Lyell put catastrophism on the road to extinction, right behind Neptunism. By the mid-nineteenth century, most scientists had accepted both the evidence that the Earth is very old[7] and that gradual, uniform, present-day processes, operating over immense spans of time, have created most of its geologic features. The influence of *Principles of Geology* went well beyond geology. Charles Darwin stepped aboard the HMS *Beagle* in 1831 with *Principles of Geology* in hand. Lyell's perspective on the immensity of geologic time prepared Darwin's mind to see evolution. "I always feel as if my books came half out of Lyell's brain," Darwin later wrote. "I have always thought that the great merit of the *Principles* was that it altered the whole tone of one's mind, and therefore that, when seeing a thing never seen by Lyell, one yet saw it through his eyes."

If anything, Lyell did his work too well. Important discoveries in geology since Lyell have stumbled over a too-strong commitment to gradualism and uniformity of earth processes. Today's Earth, it turns out, gives us little insight into some events of its past. For instance, we know that large asteroids have smacked into the planet many times in the past—with devastating effects on life and climate. These events are rare and catastrophic in the extreme. We cannot observe or measure them directly today because—luckily—a gargantuan asteroid impact has never happened in human history. As another example, evidence suggests that about 700 million years ago, glacial ice covered the planet from poles to equator, effectively turning the Earth into a giant snowball. Such a planet certainly bore little resemblance to today's Earth, and we can't understand this "Snowball Earth" of 700 million years ago by looking at what today's ice caps and glaciers are doing. Geology has had to grow beyond the teachings of Hutton and Lyell, but that in no way diminishes their stature as founders of the science—just as Einstein will always stand as a giant in physics, even as that science has grown beyond his contributions.

THIS, THEN, was the state of geologic understanding in the mid-nineteenth century, as the emigrants stepped off onto the west bank of the Missouri River, bound for California. The gradualist, uniformitarian thinkers—the minions of Hutton and Lyell—were in ascendance. Neptunists were vanquished, catastrophists in retreat. Biblical literalists were stubborn—and remain so today. Some emigrants knew about Lyell. Many recognized the evidence for the great age of the Earth and understood that gradual geologic processes, operating over immense spans of time, could produce dramatic landscapes. For instance, Riley Root in

7. Just how old—4.55 billion years—would not be revealed until the twentieth century with the discovery of radioactive decay and its application to measuring the ages of rocks.

1848, pondering the Missouri River's deeply entrenched valley, theorized that the river "carries onward to the ocean more [sediment] than it receives, and thereby causes a lowering of its bed, though not visible for ages, yet gradually and slowly has it worn away the earth to its present conditions." Likewise, James Clyman recognized that the fantastically eroded rock formations along the North Platte Valley were the work of the same modest streams "still in active operation."

Other emigrants, not surprisingly, had no conception of the transforming power of slow geologic processes acting over vast spans of time. Many viewed the world through a biblical lens. For instance, at Devils Gate (a spectacular chasm along the Sweetwater River that we'll visit in chapter 5), 1860 emigrant Martha Missouri Moore stood "in awe of Him Who tore asunder the mountains," while John Edwin Banks gazed at the eroded rock formations along the North Platte Valley and decided that they had been "torn by the rushing flood."

With the hindsight of more than a century and a half of scientific advance, we can see big gaps in the emigrants' understanding of geologic processes. They could not know the reason why mountains rise or why rocks deform, for example. The root cause of these things lies in plate tectonics—a discovery that lay more than a century in the future. They were baffled by the Rockies, where miles of rock have been heaved skyward and twisted up like putty. For many of them, volcanic activity seemed like the only force strong enough to do such work (a view consistent with Hutton's idea that the Earth's internal heat lifted up mountains). They therefore called upon volcanism repeatedly to explain what they saw— even in places where there are no volcanic rocks in sight. "A volcanic eruption must have rent the earth asunder," Dan Gelwicks decided at Ash Hollow on the Great Plains (where no volcano has erupted for well over 1 billion years). Elizabeth Dixon Smith thought that the eroded rock formations of the North Platte Valley had been "thrown up by volcanic eruptions." Likewise, James Pritchard described tilted sandstone layers along the flanks of the Laramie Range as having been "thrown up to a great hight by the upheaveing of volcanoe eruptions," while James Clyman proposed that the mountain ridges of the Basin and Range showed "their volcanic origin by their standing in the form of waves of the ocean." The emigrants took what they saw in the West and fit it into the framework of geologic understanding that prevailed in their time. We do the same thing today.

ASCENDING THE PLAINS

Most westbound emigrants were both religious and practical folk, and two holy trinities mattered a lot to them. One was always with them, they hoped, but the other they had to find. This was the trinity of grass, wood, and water. "We had good *grass, wood & water* [original italics] & good spirits," Wakeman Bryarly wrote one night along the Humboldt River, "& with these luxuries, after a pleasant smoke we retired to dream of home & love." Wood was the least essential of the three. In its absence, buffalo chips (dried dung patties) on the Great Plains or the ubiquitous sagebrush and greasewood farther west made serviceable cooking fires. But grass and water were absolute necessities. River bottomlands were the most reliable sources of grass and water, and often wood. River valleys also offered smooth slopes for overland travel. River valleys thus became the main corridors of westward migration.

The first great river corridor west was the Platte River. Fed by tributaries rooted in the snowy ranges of Colorado and Wyoming, the two forks of the Platte, North and South, converge in present western Nebraska and continue east to the Missouri River, joining it below present Omaha. The Platte's wide valley and gentle gradient formed a natural road across the Great Plains, and trails west converged on the Platte like roots on a tree. Emigrants followed the Platte, the North Platte, and the Sweetwater rivers all the way to South Pass—nearly halfway to California, and by far the longest stretch of the overland trail along a single river drainage system.

Most emigrants set off from one of three Missouri River towns: Independence, St. Joseph, or Council Bluffs. Independence and St. Joseph were the main jumping-off points during the 1840s, with St. Joseph taking more of the traffic in the years leading up to the California gold rush. Margaret Frink described the scene at St. Joseph on April 23, 1850:

> The whole country around the town is filled with encampments of California emigrants. This is the head of the emigration at the present time. They have gathered here from the far east and south, to fit out and make final preparations for launching out on the great plains, on the other side of the Missouri River. . . . [T]housands are waiting patiently for the grass to grow, as that will be the only feed for their stock, after crossing to the west side and getting into Indian country.

Both Independence and St. Joseph lay well south and east of the Platte River. To reach the Platte from Independence (near present Kansas City), the emigrants headed upstream along the valleys of the Kansas and Little Blue rivers, arriving on the south bank of the Platte near Fort Kearny. From St. Joseph (about 60 miles upriver from Independence), they headed west to the Little Blue River to join the trail coming up from Independence, likewise arriving on the south side of the Platte near Fort Kearny. From there everyone rolled west along the south bank of the Platte. This south side trail is the original Oregon-California Emigrant Trail. It served the early years of emigration to Oregon and handled most of the traffic during the gold rush years. The south-side trail had a singular disadvantage, though—it forced emigrants to cross both the South Platte and the North Platte rivers somewhere along the way. River crossings on the Oregon-California Trail were often hazardous affairs. Hundreds drowned—bitter irony on a journey through some of the most arid land in the western hemisphere. Countless others lost property when wagons flipped or livestock drowned.

During the 1850s and 1860s, as consistent steamboat service moved farther north up the Missouri River, Independence and St. Joseph ceded more and more emigrant traffic to Council Bluffs, located about 120 miles upriver from St. Joseph near present Omaha. Council Bluffs was both closer to the Platte River and lay north of the Platte's confluence with the Missouri. The trail west from Council Bluffs—called the Council Bluffs Road—therefore took travelers to the north side of the Platte. The Latter-day Saints used the Council Bluffs Road during their exodus to the Salt Lake Valley during 1847; hence the other name for this route—the Mormon Trail.

Wherever they set out from—Independence, St. Joseph, Council Bluffs, or several other jumping-off points on the Missouri—the emigrants had the company of the muddy Platte River for much of their journey across the Great Plains. During the peak emigration years, they also had the company of many, many other emigrants. Margaret Frink described the scene along the Platte Valley on May 20, 1850:

> [The] roads were thickly crowded with emigrants. It was a grand spectacle when we came, for the first time, in view of the vast emigration, slowly winding its way westward over the broad plain [of the Platte Valley]. . . . It seemed to me that I had never seen so many human beings in all my life before. . . . I thought, in my excitement, that if one-tenth of these teams and these people got ahead of us, there would be nothing left for us in California worth picking up.

THE PLATTE is silty, shallow, and swift. Its multiple channels branch out across a swath of riverbed that in emigrant days was often more than a mile wide, though rarely more than a few feet deep. It is a classic braided river, a pattern formed where rivers are so choked with sand that the water splits up into multiple channels around sand islands, forming a pattern like sloppily braided hair. The islands are elongate, ranging from a few yards to a few miles long, and aligned with the current. The sand is pieces of the Rocky Mountains ground up twice over. Most of it comes from eroding sedimentary layers on the high western Great Plains, which in turn formed from the eroded debris of the Rockies themselves.

The Platte's bottomlands today support thick stands of mature trees — far more than when the emigrants passed through. Major Osbourne Cross wrote in 1849 that the Platte possessed "more islands half covered with useless timber than any other stream of its size in the country." By "useless timber," he doubtless meant small and brushy. Trees had a harder time taking hold along the Platte in emigrant days because prairie fires periodically scorched the bottomlands, and because spring floods scoured away seedlings. Today the fires are suppressed, and dams and irrigation have tamed the river's flow.

Navigation has always been impossible on the Platte. Boats of even the shallowest draft continually run aground on sandbars. In emigrant days, fur traders sometimes tried to move their wares downstream on shallow draft boats, as Edwin Bryant saw on his 1846 journey.

> [The fur traders] were navigating two "Mackinaw boats" loaded with buffalo skins, and were bound for the nearest port on the Missouri. [One] stated that they had met with continual obstructions and difficulties on their voyage from its commencement, owing to the lowness of the water, although their boats, when loaded, drew but fifteen inches. They had at length found it impossible to proceed. . . . Their intention now was to procure wagons if they could, and wheel their cargo into the settlements.

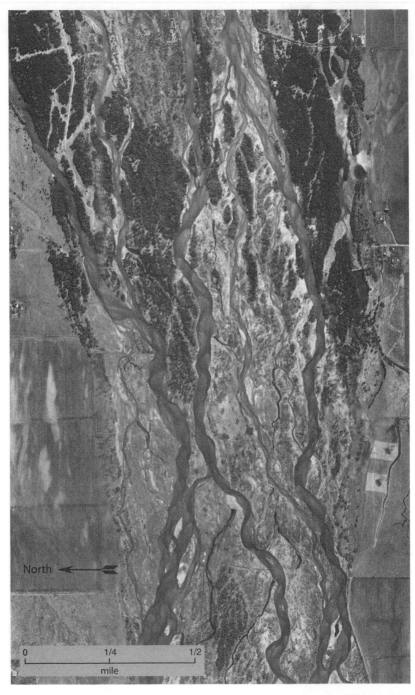

North ←

0   1/4   1/2
mile

3.1 Aerial view of the Platte River near Kearny, Nebraska, showing its wide floodplain crossed by multiple braided channels.

The muddy Platte spawned well-worn clichés: so muddy it flowed bottom-side up; so muddy you could chew it; so muddy you could plow it; a mile wide and an inch deep. The muddy water was unpleasant and even sickening to drink. Amelia Stewart Knight described its effects during her 1853 journey. "Still in camp, my husband and myself being sick (caused we suppose by Drinking the river water, as it looks more like dirty suds, than anything else) we concluded to stay in camp, and each take a vomit, which we did, and are much better. . . ." The water could be improved a bit by letting the mud settle out in a pail. "Water poor white with clay of which the Platte and the Mo. [Missouri] are alike," Ameila Hadley remarked. "But by taking a pail full, and by putting in a little alum, it will settle in a short time."

Another trick for water along the Platte was to dig a hole a few feet deep near the river and let clean groundwater seep into it. Springs gave the best water of all. On hot days along the Platte, a cool spring was next door to heaven, as Edwin Bryant explained:

> We encamped this afternoon about a mile from the junction of the north and south forks of the Platte, near a spring of cold pure water, than which to the weary and thirsty traveler in this region nothing can be more grateful and luxurious. Nature in this region is parsimonious in the distribution of such bounties, and consequently when met with their value is priceless to those who have suffered through a long day's march under a burning sun, and whose throats are parched with dust and heat.

If the Platte's water was bad, at least it was always there. Many emigrants would wistfully remember the Platte on waterless stretches of the trail in months ahead. They would also remember the gorgeous Platte Valley, paved with succulent young grass, wildflowers, and blooming prickly pear. The road was smooth and level and the wagon trains made good time, sometimes more than 20 miles a day. Without question, the Platte River Road was the easiest stretch of the journey west.

Several weeks out and several hundred miles out from the Missouri, the emigrants began to sense that they were in real western wilderness. Most had lived their entire lives among the verdant forests and cultivated fields of the East and Midwest. Climbing the low bluffs that flank the Platte Valley, they gazed with astonishment at an ocean of grass meeting an ocean of sky in every direction. The only visible trees, mostly cottonwoods, closely hugged the river. "Our chief inconvenience here is the want of firewood," Margaret Frink explained. "There being no timber except the cottonwoods and willows along the river, it often happens that we find hardly enough to cook our meals. But Mr. Frink adopted the plan of gathering up all the fragments we found and hauling them until time of need."

"Wood is now very scarce, but 'Buffalo chips' are excellent—they kindle quick and retain heat surprisingly," Tamsen Donner reported near the forks of the Platte. As trees dwindled westward, buffalo chips became the primary fuel for campfires. Many emigrants grew quite fond of the ubiquitous platter-sized splatters. "Burning with a lively blaze and producing a strong heat . . . an excellent substitute for wood," declared one. "They emit a delicate perfume," another enthused, adding, "It takes an average of about five bushels to cook supper and breakfast for twelve persons." "They are so great a matter of convenience that we forget their origin," another wrote. A morning without dry buffalo chips was a morning without joy. "The buffalo chips being too wet to ignite, we were forced to leave our encampment without our coffee, a great deprivation under present circumstances." Buffalo chips provided amusement too. "It is the duty of the cooks on arriving at a camping place . . . to sally forth and collect chips for cooking. . . . [They] jump from the wagons, gunnysack in hand, and make a grand rush for the largest and driest chips. The contest is spirited and always fun-provoking."

Buffalo chips heralded buffalo, the feature attraction of travel across the Great Plains. Back at home, the emigrants had heard fantastic stories about the vast plains buffalo herds. Now they yearned to see for themselves. "We are beginning now to look for buffalo, with great curiosity and interest," Edwin Bryant noted as his party entered buffalo country. "Every dark object descried upon the horizon is keenly scrutinized, and manufactured into one of those quadrupeds, if its shape, color, and proportions, can be tortured into the slightest resemblance."

When the first thrilling calls of "Buffalo ahead!" echoed up and down the wagon line, everything came to a chaotic halt. The men grabbed rifles, mounted horses, and charged off, vying to be first on the hunt. It took considerable marksmanship to fire from a galloping horse and plant the ball in a fatal place. Bryant described the strategy:

> Experienced hunters aim to shoot them in the lungs or the spine. From the skull the ball rebounds, flattened as from a rock or a surface of iron, and has usually no other effect upon the animal than to increase its speed. A wound in the spine brings them to the ground instantly, and after a wound in the lungs their career is soon suspended from difficulty of breathing. They usually sink, rather than fall, upon their knees and haunches, and in that position remain until they are dead, rarely rolling upon their backs.

Now and then, the buffalo got the upper hand, as J. Goldsborough Bruff explained.

Cooking on the plains
(Buffalo-chip fuel.)

3.2 Cooking over a buffalo-chip campfire on the Great Plains, as sketched by J. Goldsborough Bruff in June 1849. Reproduced by permission of the Huntington Library, San Marino, California.

The casualties of buffalo hunting are very common. Men charg'd by wounded bulls, unhorsed, and many badly hurt—the horses generally running off with the band of buffaloes, for the Indians to pick up hereafter. Lots of rifles and pistols lost, as well as horses: and many a poor fellow, after a hard day's hunt, on an empty stomach, unhorsed some distance from camp, has a long and tiresome walk, after night, to his own, or the nearest camp he can make. And some have been lost for days, at the imminent risk of their lives.

Notwithstanding such mishaps, buffalo hunts made for high times on the high plains, as thick buffalo steaks sizzled over buffalo-chip campfires. "The sweetest and tenderest meat I have ever eaten," William Swain decided. Edwin Bryant thought the "beef from a young fat heifer or cow (and many of them are very fat) superior to our best beef. The unctuous and juicy substances of the flesh are distributed through all the muscular fibres and membranes in a manner and an abundance highly agreeable to the eye and delightful to the palate of the epicure."

The sheer multitudes of buffalo astonished everyone—herds of thousands stretching to the horizon. Yet some spoke with prescience about the excessive slaughter and its implications. "Not less than fifty buffalo were slaughtered this morning, whereas not three in all were used," wrote a disgusted Lorenzo Sawyer in May 1850. "Such wanton destruction of buffalo, the main dependence of the Indians for food, is certainly reprehensible. But, the desire of the emigrant of engaging once at least in a buffalo chase can scarcely be repressed."

They proceeded on, ever west, up the gentle valley of the Platte. Flocks of songbirds whirled by on rushing wings. Wolves loped by during the day and howled nearby at night. Prairie dog towns lay scattered across the landscape like subterranean cities. The prairie dogs "come out of their burrows by thousands, and standing perfectly erect on their hind feet, impudently bark with their sharp voices at the passing multitude." Forty-niner A. J. McCall was particularly intrigued by these plentiful rodents.

> The prairie dog is of a sprightly mercurial nature, quick, sensitive and somewhat petulant. He is entirely gregarious, living in large communities, sometimes of several acres in extent. . . . They are constantly barking and bobbing about, like other village gossips, making a din and clatter like the assembled wisdom of the nation, as if the affairs of the world were in their keeping. . . . They are very wary and dodge into their holes at the approach of intruders.

McCall's fondness for prairie dogs was also epicurean. "One of our hunters killed a brace of dogs, and we had them stewed for our dinner and found them quite palatable—in fact they were excellent. It matters not what you call a thing—good or bad—the proof is in the eating."

Antelope appeared in ever-greater numbers westward but were hard to bring in for meat. Hunters would approach with utmost stealth, ready to give chase on their fleetest horses. They usually turned back disappointed. Edwin Bryant described a typical hunt:

> The antelopes did not discover us until we had approached within the distance of half a mile. They then raised their heads, and looking towards us an instant, fled almost with the fleetness of the wind. I never saw an animal that could run with the apparent ease, speed, and grace of these. They seem to fly, or skim over the ground, so bounding and buoyant are their strides, and so bird-like their progress.

Some hunters had better success with antelope using unconventional methods, as Dr. Charles Parke discovered.

> You cannot slip up on an antelope, but you can excite their curiosity, entice them up to you, and shoot them. This is a simple procedure. All that is necessary is to hide behind a "sage bush," draw your ramrod from your rifle, and fasten a *red* handkerchief or shirt on the end of it, and move the red object back and forth. As soon as it is seen by the antelope,

3.3 Wagon train camp along the Platte River, as sketched by J. Goldsborough Bruff on July 1, 1849. Reproduced by permission of the Huntington Library, San Marino, California.

he will circle round it gradually approaching nearer and nearer until within good shooting range.

About 300 miles west of the Missouri River, the emigrants arrived at the forks of the Platte, where the South Platte and North Platte rivers converge. Their journey west had taken them across the heart of the Great Plains.

∗

Crossing the North American continent in the mid-nineteenth century committed a person to a level of separation from home and loved ones that is hard for us to fathom today. To cross the continent back then often meant years of separation from family and friends, with the cords of emotion sustained only by months-old letters. "Dear wife, my heart bleeds within me to think of starting on the west plains to be gone so long from you," forty-niner Joshua Sullivan pined. Agnes Stewart, crossing in 1853, wrote tearfully in her diary of a dear friend, "I miss you more than I can find words to express. I do not wish to forget you but your memory is painful to me. I will see you again I will if I am ever able I will go back." Separation was no easier for those left behind. "O!! how I want to see you," Sabrina Swain wrote to her absent forty-niner husband. "William, if I could see you this morning, I would hug and kiss you till you would blush." Jet travel and interstate highways make it easy for us to forget how dauntingly *big* the North American continent is—nearly 3,000 miles across east to west, more north to south. Where did this colossal slab of rock come from? The answer, it turns out, lies deep below the Great Plains.

Traveling the Great Plains today, you could be forgiven for yawning at the landscape. Fields of wheat, corn, and soybeans stretch to a flat horizon. Rock stays well hidden beneath soil, crops, and grass. Where it appears in road-cuts and the eroded walls of river valleys, it presents a layer cake of strata—horizontal beds of sandstone, siltstone, shale, limestone, and occasionally volcanic ash—lying as flat as the nearby fields. The layers tell of the comings and goings of shallow oceans, deltas, rivers, lakes, and swamps, and of volcanoes erupting far to the west.

The middle of North America has gone about this quiet business of laying down horizontal strata for most of the last 500 million years, since early in Paleozoic time.[1] Meanwhile, events that were more dramatic unfolded elsewhere. Himalayan-sized mountains rose along a line from Nova Scotia to Georgia, squeezed up as North America crunched ponderously into Europe and Africa, sandwiching assorted small continents and island archipelagoes in between. Erosion has since brought these heights down to the subdued nubs that we call the Appalachians. North America then tore away from Europe and Africa and headed west. Mountain building shifted to the west side of the continent, and rock climbed skyward all the way from Alaska to Panama, making the North American Cordillera. Meanwhile the center of the continent, sheltered from these upheavals at its edges, went on accumulating layer upon layer of strata.

But this is recent history, like starting U.S. history with Watergate and ignoring what came before. The exposed rock of the Great Plains encompasses just the latest few percent of geologic time. The middle of North America seems staid today, but it was a lot wilder when it was young. In

---

1. The Geologic Time Scale on page xii gives the time ranges of geologic periods and major events in North America's geologic history.

fact, 2 billion years ago there was no middle, because there was no North America. The continent lay scattered in pieces, still to be assembled. Deep below the Great Plains lays rock that reveals nothing less than the birth of the continent.

THE EARTH'S continents did not form along with the Earth itself. They came later. According to current theory and evidence, continents form and grow by accretion at their edges. Small bits of crust raft together like pieces of floating wood colliding in a stream. They stick together, piece by piece, amassing over time into a sizable continent. This happens because the Earth's surface is in continual motion, driven by the convective churning of its hot interior. The motion pushes bits of crust around on the Earth's surface so that they eventually collide and join.

In plate tectonic theory—the unifying concept of geologic thought— the Earth's outer surface is divided like a suit of armor into several dozen separate, moving rock plates. The plates range 50 to 100 miles thick and vary widely in breadth and shape. The Pacific Plate, the largest, covers one-fifth of the Earth's surface. The smallest plates would fit inside of Texas with room to spare. The plates slide across the face of the Earth like sheets of ice drifting on a winter pond. They move several inches each year, made mobile by a layer of hot, semi-solid rock in the mantle below. Continents are not plates, but they form parts of some plates. Continents ride within plates like yolks in fried eggs. Where the plates go, so go the continents, wandering the globe, slowly drifting together in some areas and splitting apart in others. Ocean basins form where plates separate; mountain belts form where they converge. We'll explore these processes in detail later, when we time-travel through the geologic evolution of the North American Cordillera. Right now, our focus is on an older and more fundamental process—how early plate movements built the North American continent.

The earliest plates formed perhaps 4 billion years ago, after the young, largely molten Earth had cooled enough for a solid crust to congeal. These plates were probably akin to oceanic plates today, moving about below a globe-covering ocean. Much of the Earth's most ancient rock—from the Archean eon, more than 2.5 billion years old—appears to have begun as sediment deposited in the deep ocean. It tells us there was little dry land or even shallow water on Earth early in its history. Today continental rock makes up about 40 percent of the Earth's crustal rind, and the remaining 60 percent is oceanic rock—the rock that floors the ocean basins. Four billion years ago, it appears that there was hardly any continental rock at all. The entire Earth then may have looked something like the Pacific today—nearly all deep ocean floor with a few scattered volcanic islands

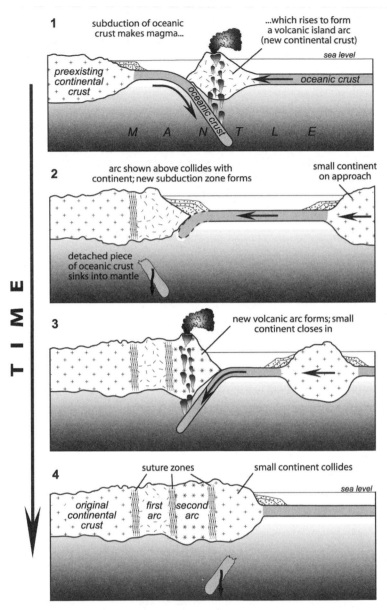

1 subduction of oceanic crust makes magma... ...which rises to form a volcanic island arc (new continental crust)

sea level

preexisting continental crust

oceanic crust

M A N T L E

2 arc shown above collides with continent; new subduction zone forms — small continent on approach

detached piece of oceanic crust sinks into mantle

3 new volcanic arc forms; small continent closes in

4 suture zones — small continent collides

sea level

original continental crust — first arc — second arc

TIME

3.4 The Earth's continents appear to be mosaics of crustal pieces assembled over time by plate movements. Volcanic island arcs are the starting points for continental growth (1). Continents grow as volcanic arcs collide along their edges and suture (2). Volcanic arcs that grow along the edges of continents also add new rock to the continent (3), as do collisions between existing continental blocks (4).

poking up. Somehow, large continental landmasses grew from this oceanic world. In other words, somehow the Earth manufactured colossal volumes of continental rock. How?

As Charles Lyell taught us, if we want to figure out how the Earth did something in the past, we can begin by looking at how it does that same thing today. The Earth makes new continental rock today in volcanic island archipelagoes—lines of volcanic islands like the Marianas, the Philippines, and the Aleutians. These archipelagoes sprout from the deep seafloor where two oceanic plates converge. The Mariana Islands, for instance, arise where the Pacific Plate, sliding northwest, converges with and dives below the adjacent Philippine Plate. The six-mile-deep Mariana Trench marks the subduction zone—the line where the Pacific Plate bends down underneath the Philippine Plate and plunges into the Earth's mantle, carrying with it quadrillions of tons of sopping wet seabed. Once the subducting plate reaches about 80 to 100 miles depth in the mantle, the intensifying heat drives trapped seawater out of the plate and into the surrounding mantle. Water has a curious effect on hot rock—it lowers its melting point. Catalyzed by water, the hot mantle just above the downplunging plate begins to melt, but not all at once. Minerals rich in lighter elements—elements like oxygen, silicon, sodium, and aluminum—melt out more readily than others. Like chocolate chips melting out of a cookie left in the sun, these more easily melted components ooze out to form magma—molten rock—leaving the remaining rock behind.

This newly born magma, enriched in light elements, then punches and melts it way up toward the Earth's surface, evolving chemically along the way. Just as the human contents of a New York City subway changes as the train passes from Harlem to the Upper East Side, so the chemical content of magma changes on its long trip up from the mantle to the Earth's surface. Lighter chemical elements (such as oxygen and silicon) get on the magma train, while heavier ones (such as iron and magnesium) get off by forming minerals that crystallize and settle out. Once the magma reaches the surface, it bursts out to form the spectacular volcanoes of the Mariana Islands. Much of the magma doesn't reach the surface. It solidifies several miles underground instead, forming a foundation for the volcanic edifices. But here is the astonishing thing—the rock that forms from this magma is not at all like the mantle from which it came. This magma, enriched with light elements and depleted of heavier ones, makes *new* types of rock—*continental* types of rock. Where it erupts, it makes andesite lava flows and volcanic tuff. Where it congeals underground, it forms diorite, granodiorite, and granite. These are the rocks of continents, not of ocean floor or mantle. The Mariana Islands are continental crust, freshly minted.

Volcanic archipelagoes that arise this way—from the subduction of oceanic plates—are called volcanic island arcs, so-named because the lines of volcanic islands describe curves that run parallel to the curving lines of oceanic trenches. The Aleutian Islands, the Philippines, Japan, the Tonga and Kermadec Islands, Java, Sumatra, and the Lesser Antilles (West Indies) are all examples of volcanic island arcs. Each lies behind an oceanic trench—the line where an oceanic plate bends down into the mantle. Each is built of rock that is continental in nature, born from the subduction of the adjacent oceanic plate, with seawater acting as midwife.

Current theory holds that ancient volcanic archipelagoes are the building blocks of today's continents. Put enough Marianas or Philippines together, the theory goes, and you can make a small continent—a Madagascar, a New Zealand, or a Sri Lanka (Ceylon). Put enough Madagascars together, and you can make an Australia, an Africa, or a North America. "Wait!" Charles Lyell shouts from his grave. "If you can't see the Earth doing this today, you can't know that it ever did it in the past." Rest easy, Sir Charles. It turns out the Earth *is* enlarging its continents this way today. Active continental growth is unfolding in Indonesia. Many of the islands of Indonesia are pieces of continental rock that would become a respectable continent if you could somehow sweep them together. Australia seems determined to do just that. Riding north with the ocean floor around it as part of the Indo-Australian Plate, Australia is on a collision course with Indonesia. Over the next several tens of millions of years, the islands appear destined to collect against the prow of Australia, forming new continental real estate.

All well and good, but what evidence shows that North America was assembled this way—by the sweeping together of scattered volcanic islands? For that, we turn to the basement. "Basement" is a generic term that describes the ancient crystalline rock that makes up the foundation of all of the continents. The basement is the granite, granodiorite, diorite, gneiss, schist, and greenstone you eventually hit if you drill down through the veneer of younger sedimentary and volcanic rocks that typically cover it. Over much of Canada, Ice Age glaciers have laid bare the basement by stripping away the veneer of younger rock above. Canadian geologists may be beset by bugs and bears, but they never lack basement to study. In the United States, the basement lies mostly hidden below younger rock. To study this basement, geologists collect samples from deep drill holes and employ remote sensing: measurements of subtle variations in gravity and magnetic strength over the land surface that track changes in the rock deep below. Mountains also help. For instance, the same basement rock that lays hidden thousands of feet below the Great

Plains rises to view in the Rockies, where it forms the uplifted cores of the mountains.

Numerous studies of North America's basement converge on one conclusion—the basement looks like a collided patchwork of ancient volcanic island arcs and other crustal fragments. Large irregular bodies of gneiss appear to be the metamorphosed cores of ancient volcanic island arcs. Wrapping in great curves around these are belts of metamorphosed sedimentary rocks and greenstones (metamorphosed basalt). Many of the now-metamorphosed sedimentary rocks probably began as sand and mud shed from the volcanic islands into the adjacent deep oceans. When the islands collided, the sediments were squished in between. Some of the belts of greenstone probably represent seafloor basalt trapped between the colliding island arcs.

THE NOTION that an entire continent can grow by smashups between ancient islands seems a bit fanciful, even surreal. Geologic history, like human history, often seems abstract when viewed from afar. It becomes less so if you can walk the ground where great events unfolded. I'm sure this is what calls so many people to places like Gettysburg. Fifty thousand men fell in three July days of 1863 at the Battle of Gettysburg, in a close fight whose outcome kept the nation united. This is the central fact of Gettysburg. Lincoln knew it when he gave his greatest speech, honoring the Union dead as men "who here gave their lives that this nation might live." [2] But I never understood Lincoln's meaning until I *saw* Gettysburg. Only when I had walked the ground did I see how narrowly the Union army had escaped disaster. Random hills and ridges, combining with bits of stunning luck and bravery, tipped the battle ever so delicately toward Union victory. In the context of the war at that moment, a Confederate triumph at Gettysburg could well have cascaded into overall Southern victory in the war, and today, rather than a single United States, we might be a balkanized collection of bickering republics. Many accounts of Gettysburg spin this virtual history scenario—but I never believed it until the day I stood on Little Round Top.

Geologic history has its great battlefields too, and to understand their meaning it likewise helps to walk the ground. If we are to believe that North America grew from a series of colossal smashups between ancient crustal fragments, then we should go look at a place where that happened—a place where two big pieces went *crunch*. We can do that along the Cheyenne Belt of southeastern Wyoming.

2. From the Gettysburg Address, read by Lincoln on November 19, 1863, at the dedication of the national cemetery for Union soldiers killed at the battle.

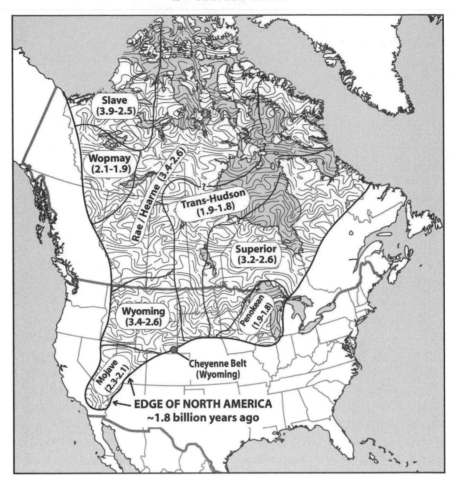

3.5 The basement rock of North America shows that the continent was assembled by collisions between large blocks of crust. The outer line shows where the edge of North America would have been approximately 1.8 billion years ago. North America at that time was roughly half the size it is today and consisted of about eight distinct crustal blocks (the geologic provinces named above) that had previously collided and sutured. The numbers give the age (in billions of years before present) of the basement rocks within each province. Note the location of the Cheyenne Belt—an exposed suture line in the mountains of southeastern Wyoming that represents the contact between pre-1.8-billion-year-old and post-1.8-billion-year-old continental basement.

The Cheyenne Belt is a band of intensely sheared and shattered basement rock, a few miles wide, that slices northeast through Wyoming's Sierra Madre, Medicine Bow, and Laramie ranges. Here, about 1.77 billion years ago, two gigantic crustal blocks collided to assemble this portion of North America. The basement rocks on opposite sides of the belt are utterly different; their juxtaposition is as geologically incongruous

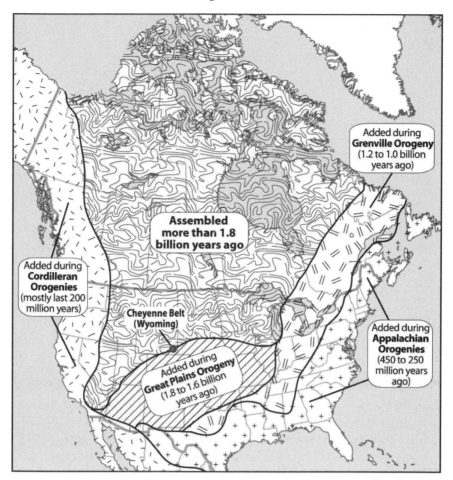

3.6 Since 1.8 billion years ago, North America has continued to grow by the addition of more blocks of crust around its edges. Island arcs and small continents have rafted up against the continent and collided, heaving up mountains in the process. Each major episode of collision and mountain building is called an orogeny. The Great Plains Orogeny was the first major episode of continental growth after 1.8 billion years ago. It created the suture zone along the Cheyenne Belt and assembled the continental basement below the Great Plains. The Cordilleran Orogenies—the youngest mountain-building episodes in North America—are discussed in detail in later chapters.

as finding parrots living beside polar bears. North of the Cheyenne Belt, the basement is made of 2.6- to 3.4-billion-year-old rock of the Wyoming Province, some of the most ancient rock in North America (mostly granite, sandstone, and limestone, all deeply metamorphosed). South of the belt, the basement rock (mostly metamorphosed volcanic lavas and sediments) is nearly *1 billion years younger*. A yawning abyss of time divides these neighboring rocks—and this is what it tells us: Step back in time to 1.8 billion years ago, and stand just north of the Cheyenne Belt in what is now southeastern Wyoming. You stand on the southern edge of proto-North America. To the north lies the partially assembled continent, made up of freshly sutured volcanic island arcs and other crustal blocks. To the south, you gaze across abyssal blue ocean—no continent, no land. Colorado, Nebraska, Kansas, Texas, Mexico, and everything else south of the Cheyenne Belt do not yet exist.

But the pieces are coming. Out in the ocean to the south, a volcanic island arc—a Philippine-like archipelago—is slowly rafting in, ferried toward you by the oceanic plate on which it rides. Advancing a few inches per year, the archipelago crunches against the edge of the continent at your feet and sutures.[3] (The collision zone is today's Cheyenne Belt, and the accreted archipelago now makes up the basement of southeastern Wyoming, northeastern Colorado, and much of Nebraska.) Beyond it, other volcanic archipelagoes are cued up for arrival. One by one, these strips of island real estate collect against the backstop of the Cheyenne Belt. From 1.8 to 1.6 billion years ago, these serial collisions assemble most of the basement below the United States, including that which lies below the Great Plains.

The force of these ancient collisions shoved up mountains across what is now the flat middle of North America. Back then, the Great Plains looked like the Alps. Erosion has long since brought those heights down, although their memory lives on in the form of metamorphic minerals that form only under mountainous weights. Mountains grow where plates collide. The process is called orogenesis, and the events are known as orogenies. The continent-building and mountain- upheaving episode that we've just explored is called the Great Plains Orogeny. To the west, we'll visit rocks and landscapes racked by younger orogenies: Nevadan Orogeny, Sevier Orogeny, Laramide Orogeny, Basin and Range Orogeny, and other lesser upheavals.

AFTER THE Great Plains Orogeny, the history of the Great Plains entered a new phase. Erosion beveled everything down to a low-lying plain that became the stable center of the growing continent. Island arc collisions, mountain building, and continental growth continued at the edges of the continent, leaving the center behind. Between about 1.2 and 1 billion years ago, North America nearly split in two, north from south, along a rift zone extending from Kansas to the Great Lakes. As the continent tore

3. This collision took place from 1.79 to 1.75 billion years ago, based on radiometric dating of zircon minerals in Cheyenne Belt rocks (Houston 1993).

asunder, basalt lava welled up and congealed like blood in a gaping 1,000-mile-long wound. So much basalt erupted along this deep basement rift, called the Mid-Continent Rift, that it changes the gravity. Things weigh a bit more over the Mid-Continent Rift because of the dense basalt several thousand feet down. (Basalt's high iron and magnesium content makes it denser than most continental rock, giving it a stronger gravitational pull.) You can't pick up the higher gravity on a Kansas City bathroom scale, but you can see it easily on a gravimeter—an instrument capable of precise gravity measurements. The rifting failed, of course; had it not, North America would today be split into northern and southern halves, perhaps divided by hundreds of miles of ocean—a Confederate's wet dream. But the forces of union won and the continent stayed whole.

Some 500 million years ago, near the start of the Paleozoic era, shallow seas began repeatedly flooding the interior of North America, sweeping across what would eventually become the Great Plains. The seas draped the naked basement with blankets of sandstone, shale, and limestone. Between invasions of the sea, more sedimentary layers piled up in rivers, lakes, dunes, swamps, and deltas. Up and up the layers stacked, several thousand feet thick above the basement in future Kansas and Nebraska.

The birth of the modern Great Plains is linked to the uplift of the Rocky Mountain Foreland Ranges, which began to shoulder upward across future Colorado and Wyoming about 75 million years ago. As the mountains rose, they shed rock debris eastward into future Kansas and Nebraska, expelling the last of the great interior seas about 65 million years ago. The eroded debris of the young Rockies—layers of gravel, sand, and silt—piled up thickest next to the mountains and thinned to the east. The surface of this east-thinning debris wedge forms today's Great Plains, which rise from about 1,000 feet elevation on the Missouri River to about 5,000 feet at the toes of the Rockies.

THIS STORY of the heartland—the center of North America—covers some 4 billion years, or about 85 percent of the span of Earth's existence. Geologic time, like astronomical light-years, embraces such stupendously large intervals that it denies comprehension at an intuitive level. Years in geologic time are like coins on Wall Street. Wall Street doesn't deal in nickels or dimes, and geology likewise doesn't fuss about decades or centuries. It is no crime in this science to say "about 120 million years ago" when you mean 119 million—even though that 1-million-year rounding error is five times longer than human beings have existed on the planet.

If inches were years, then two or three walking steps would take in your life span. The distance of a short fly ball to the outfield would take in the time since the days of Jesus, and a well-hit home run would encompass the time since the pharaoh Khufu (Cheops to the Greeks) laid plans for his pyramid. These kinds of distances represent the scale of human history. To ramp the analogy up to geologic time, consider this: the Earth is about 24,000 miles in circumference at the equator, which equals 1.52 billion inches. The Great Plains Orogeny hoisted mountains across future Kansas and Nebraska between 1.8 and 1.6 billion years ago—an event more ancient in years than the Earth is fat in inches. The oldest basement rocks in the Wyoming Province north of the Cheyenne Belt are twice as old—two times around the Earth, in inches. The Earth itself is about 4.55 billion years old—three times around the planet, and then some, in inches.[4] That's a lot of history, and much of it lies on the road ahead.

4. Put another way, 4.55 billion is roughly the number of seconds in 145 years. The estimate of 4.55 billion years for the age of the Earth comes from radiometric dating. Like a ticking clock, certain elements in rock change into new types of elements at a precise rate (a process called radioactive decay). Uranium, for instance, decays into lead over time. The more time goes by, the less uranium and the more lead in the rock. The oldest rocks found so far (in Canada's Northwest Territories) are 3.96 billion years old, and grains of zircon minerals eroded from now-vanished rock in Australia date to 4.2 billion years. Since erosion and moving tectonic plates are effective destroyers of rock on Earth, the Earth must be older than its oldest rocks. The oldest moon rocks and meteorites consistently date between 4.5 and 4.6 billion years old. Since we assume that the Earth formed along with the rest of the solar system—including the moon and meteorites—we use 4.55 billion years as our current best estimate for the age of the Earth.

EXHUMED MOUNTAINS
AND HUNGRY RIVERS

On the western Great Plains a few miles west of the Platte forks, the emigrants faced a fearsome ordeal: the fording of the South Platte River. Margaret Frink pulled up to the ford on May 28, 1850, and found the river "fearful to look at—rushing and boiling and yellow with mud, a mile wide, and in many places of unknown depth. The bed was of quicksand—this was the worst difficulty. But there was no way to do but to ford it."

Anyone traveling on the south side of the Platte River had to ford somewhere to continue west, because the trail to California and Oregon follows the valley of the North Platte River west of the Platte forks. Rather than fording downstream of the forks, where the river is widest, south-side emigrants followed the South Platte west for some distance past the forks to a suitable fording point. Most forded at either Lower California Crossing near present Brule, Nebraska, or Upper California Crossing 20 miles upstream near present Julesburg, Colorado.[1]

Irrigation farming today robs the South Platte of much of its flow. But the river's half-mile-wide braid plain testifies to times in the past when it

---

1. Emigrants on the north side of the Platte, on the Council Bluffs Road/Mormon Trail, had no river crossing worries at the Platte Forks. They met the North Platte at the forks and just kept rolling west on the north bank of the river. But 200 miles ahead, near Fort Laramie (chapter 5), north-side travelers had to cross the North Platte to its south side in order to continue west. Once Child's Cutoff opened in 1850, north-side travelers could stay on the north side of the North Platte the whole way.

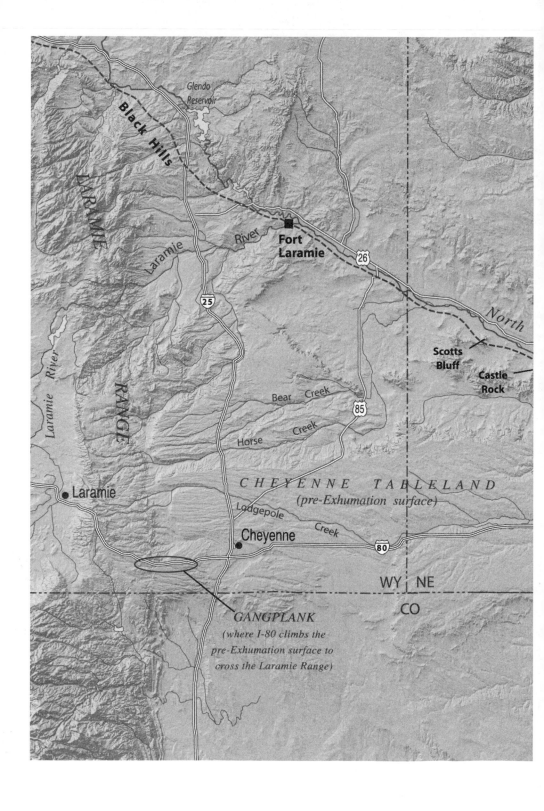

Glendo
Reservoir

Black Hills

LARAMIE

Laramie   River

**Fort
Laramie**

26

25

North

**Scotts
Bluff**

**Castle
Rock**

RANGE

Bear   Creek

85

Creek

Horse

Laramie River

CHEYENNE   TABLELAND
(pre-Exhumation surface)

• Laramie

Lodgepole

Creek

Cheyenne •

80

WY | NE

CO

GANGPLANK
*(where I-80 climbs the
pre-Exhumation surface to
cross the Laramie Range)*

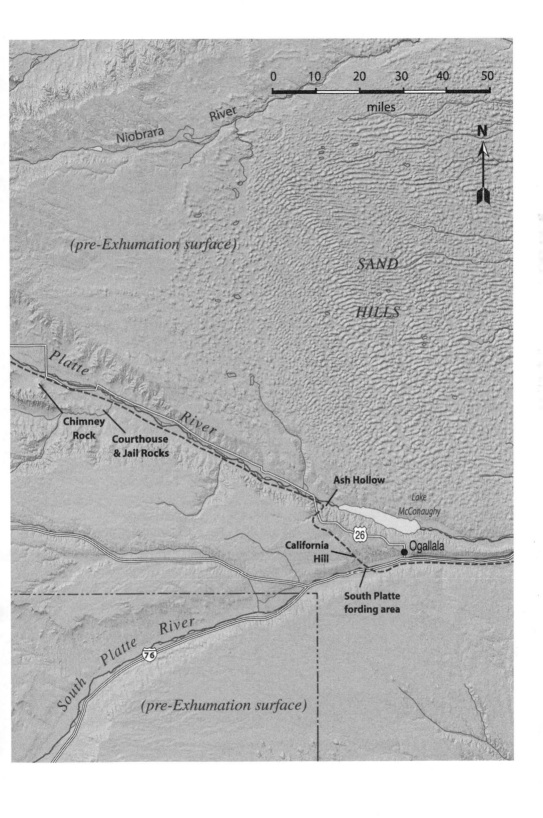

rushed east with all of the enthusiasm of a wild river. In emigrant days, reinforced by spring runoff, the South Platte could be a holy terror. At fording time, everyone stowed and secured their supplies as best they could. Some raised the wagon beds on wooden blocks to gain a few more inches of clearance above the churning waters. For extra pulling power, emigrants often double-teamed with six to eight yoke of oxen or mules on each wagon. If the route across looked uncertain, a line of men on horseback crossed first, feeling for depth, hidden holes, and quicksand. Then the drivers plunged in with the teams and wagons. There was no stopping, for as A. J. McCall explained, "When the wheels struck the quicksand the swift water washed it out from under, so that the wheels are constantly running up hill. . . . The moment the wagon or team halts, they are going down — down and would be soon entirely buried."

The swirling current and unstable footing sometimes panicked the animals, and they instinctively tried to turn downstream or go back. The drivers lashed them on. Upon gaining the far bank, the animals were unhitched from the wagons and driven back across to team up on yet more wagons. It typically took several nerve-racking hours for an entire wagon train to cross this way. Margaret Frink declared her safe arrival on the north bank "one of the happiest and most thankful moments of my life."

Once across, everyone struck out overland toward the North Platte. From Lower California Crossing, they rolled up a gentle rise today called California Hill. Countless hooves, boots, and wheels followed a single track up this hill, biting deeply into the soft loess soil.[2] Over the years the traffic excavated a cut some 20 feet wide, 60 feet long, and several feet deep at the center. It is still there today, in the heart of Nebraska's wheat and cattle country.

Plows and tractors have obliterated most traces of the original emigrant trail on the Great Plains. That's why the segment of original trail at California Hill is a must-see for anyone chasing emigrant shadows. I went there in the fading light of a stormy June evening. A stiff wind slung

---

2. Loess (rhymes with *bus*) is deposits of windblown silt. Loess covers large areas of Nebraska, South Dakota, Iowa, Missouri, Illinois, and the Columbia Plateau of Washington State. It appears as powdery, tan-colored, cliffy outcrops where exposed in ravines and road-cuts. Loess forms wherever strong winds and an ample supply of silt coincide. Much of the world's loess formed during the ice ages (the informal term for the glacial periods of the Pleistocene epoch). Advancing continental glaciers ground vast amounts of rock into silt, which meltwater streams washed out past the glacial fronts. Winds then picked up the silt and spread it across large segments of the Midwest and West. Loess-based soils drain readily and are rich in finely ground mineral matter, making them some of the most productive farmland on Earth. The Midwest and the Great Plains owe much of their high agricultural production to loess.

4.2 California Hill, about five miles west of Brule, Nebraska. Thousands of wagons ascended this hill after fording the South Platte River, and over the years they cut this cleft in the soft loess.

4.1 (*previous pages*) Shaded relief map of the western Great Plains along the route of the Oregon-California Trail. The trail ascends the valley of the North Platte River to pass around the north end of the Laramie Range. Until about 5 million years ago, sedimentary layers nearly covered the Laramie Range and other Rocky Mountain ranges and spilled thickly east across the ancient Great Plains. During the past 5 million years, east-flowing rivers like the North Platte and South Platte have been busy chewing up this sedimentary layer cake. The eroding rivers have exhumed the mountains from burial and have cut deeply into the layers on the plains. Surviving remnants of the layer cake form the famous rock monuments of the North Platte Valley—Courthouse Rock, Chimney Rock, Scotts Bluff, and others. Large sections of the layer cake still exist beyond the reach of the rivers. Interstate 80 follows one such section today—the Cheyenne Tableland, which rises like a ramp smoothly west from the plains to the summit of the Laramie Range. The uppermost part of this sedimentary ramp is called the Gangplank.

raindrops sideways while black clouds spat lightning along an advancing front to the west. June is thunderstorm and tornado season on the Great Plains, and the sky boils up dark and ugly on many afternoons. June was also the peak season of emigration across the plains. There can be no doubt that emigrants rolled up California Hill on similar stormy June days. Here creaking wheels may have rolled to a halt, the sweating oxen chuffing and tossing their heads, agitated by the flashes and cracking thunder. Men would have gathered, eyeing the advancing storm and pondering the risks. Shall we go forward and chance a lashing on the open tableland ahead? Do we burn up time and effort to go back to the sheltered bottomlands along the South Platte? Within minutes of my arrival at California Hill, the sky came down like a waterfall. Lightning painted harsh silhouettes across the gloaming plains while thunder shredded the air. It would have been a sleepless night huddled in a covered wagon. Amelia Stewart Knight (traveling in 1853 with her husband and seven children, and several months pregnant with the eighth) described what it was like to wait out such a storm:

> We had a dreadful storm of rain and hail last night and very sharp lightning. It killed two oxen for one man. We have just encamped on a large flat prairie, when the storm commenced in all its fury and in two minutes after the cattle were taken from the wagons every brute was gone out of sight, cows, calves, horses, all gone before the storm like so many wild beasts. I never saw such a storm. The wind was so high I thought it would tear the wagons to pieces. Nothing but the stoutest covers could stand it. The rain beat into the wagons so that everything was wet, in less than 2 hours the water was a foot deep all over our camp grounds. As we could have no tents pitched, all had to crowd into the wagons and sleep in wet beds with their wet clothes on, without supper.

Of all the Great Plains storms, hailstorms were the worst. Elizabeth Dixon Smith endured one on July 8, 1847. "To day we had the dredfulest hail storm that I ever witnessed. . . . [I]t tore some of their waggon covers off broke some bows [the curving wooden stays that hold up the canvas covers] and made horses and oxon run a way." Near the South Platte crossing on June 20, 1849, William Swain's party was pounded by hailstones "the size of a *walnut* to that of a *goose egg* [original italics]." The terrified livestock, cut to bleeding and "writhing with the pain inflicted by the strokes of the hail," reared and spun in their traces, upsetting wagons and breaking tongues and wheels. People hid under wagons or grabbed saddles, pails, or kettles for shelter. When the bombing ended, all gathered around to compare their "sundry bruised and gashed heads, black eyes, pounded and swollen backs, shoulders, and arms, which with a little attention from the doctor and some liniment soon became sound."

4.3 Wagons huddled under the wrath of a Great Plains thunderstorm, as portrayed by J. Goldsborough Bruff in June 1849. Reproduced by permission of the Huntington Library, San Marino, California.

Laughter spread with the emerging sun. "'No great evil without some good' was our motto," Swain wrote, "so we filled our pails and kettles with hail and had ice water the rest of the day, a luxury we little expected on this route."

Leaving California Hill, the trains rolled northwest across a high prairie called the Cheyenne Tableland. The tableland ascends smoothly west, like a great ramp, nearly to the crest of the Laramie Range more than 150 miles to the west. Arriving at the northern edge of the tableland, the emigrants spied the North Platte River in the distance—the stream that would be their near-constant companion for the next 300 miles.

The character of the Great Plains changes west of the Platte forks. The landscape becomes rougher, more eroded and gullied, and this trend continues west toward the Rocky Mountains. Creeks flowing off the Cheyenne Tableland toward the North Platte have clawed their way headward, south, into the tableland. They make fingerlike ravines that are impassably steep nearly everywhere along the northern escarpment of the tableland. Choosing the best among several bad options, the emigrants found a route down off the tableland into the valley of Ash Hollow Creek. At one spot the trail plunges nearly 200 vertical feet over a distance of a few hundred yards. "The road hangs a little past the perpendicular," Lewis Dougherty quipped at this notorious incline today called Windlass Hill.

The descent of Windlass Hill was a dangerous maneuver. There is no evidence that emigrants used actual windlasses to lower their wagons here. (They never called it Windlass Hill either—the name appeared rather mysteriously in later years and although incorrect, it stuck.) Instead, they unhitched all or part of their teams at the top of the hill, rough-locked the wagon wheels with chains, and tied ropes to the rear axles. Then groups of men dug in their heels and slowly eased out on the ropes to let the wagons down the hill. Some men worked the wagon tongues to steer; others lifted and eased the locked wheels over the many rock ledges jutting from the slope. Most parties managed to get their wagons down this way without incident. But sometimes wagons broke free, careening down to smash into splintery heaps at the bottom. Charles Scott described the mayhem on one such day:

> A general runaway and smash up at Ash Hollow, a terrific scene. Horses dashing furiously with the pieces down the hills and precipeces the noise, dust and confusion, the men shouting hallooing, the women screaming, made an impression on my memory, never to be effaced. Two horses were killed and seven disabled and unfit for service, in all about $25,000 damage done.

If your wagon had a smashup on Windlass Hill, at least the pretty vale of Ash Hollow was a good place to make repairs. There was water, wood, and shade, and often some abandoned wagons to scavenge for spare parts—as Margaret Frink explained:

> We remained in camp all day [June 2, 1850], repairing our small wagon. The hind axle was broken. Mr. Frink had seen a wagon abandoned, near the road at Ash Hollow. He went back with a man to-day, and took out the bolts and brought the hind axle and wheels to camp. It was then fitted to the small wagon in place of the old axle, and did very well.

EMERGING FROM Ash Hollow, the trains rolled on—ever west—along the south bank of the North Platte. "Its width is not so great [as the Platte]," Edwin Bryant observed, "but still it is a wide stream, with shallow and turbid water, the flavor of which is, to me, excessively disagree-able." The North Platte may taste bad, but in June its valley is dazzling. The full glory of spring paves the land with a carpet of lush grass and wildflowers. The bulbous yellow-white flowers of blooming yuccas pep-per the landscape. Thunderstorms regularly scrub the air clean, leaving it cool and pungent with the smell of wet grass and sagebrush. Much of the valley today is a patchwork of irrigated fields and fenced cattle ranges. Up on the valley flanks, though, far from the river and on land too steep to irrigate, you can still sense the wild high plains—the vanished buffalo grasslands.

Several days' travel beyond Ash Hollow, the emigrants rolled into an alien landscape. Huge rock formations stepped out, one by one, from the high bluffs on the south side of the North Platte Valley. First to appear were the twin monoliths of Courthouse Rock and Jail Rock. Then in 20 more miles, Chimney Rock—the most sublime—then Castle Rock, and finally Scotts Bluff, the largest and grandest monument of all. After weeks of travel over the monotonous prairie, the emigrants could not get enough of these craggy landforms. They clambered up and over the rocks, took in the grand vistas, and scratched their names into the soft strata.

> No conception can be formed of the magnitude of this grand work of nature [Chimney Rock] until you stand at its base & look up. If a man does not feel like an insect then I don't know when he should. (Elisha Perkins, June 27, 1849)

> [This morning a] line of pale and wintry light behind the stupendous ruins . . . served to define their innumerable shapes, their colossal gran-deur, and their gloomy and mouldering magnificence. (Edwin Bryant, June 22, 1846)

> The soul must be cold that can calmly view these scenes. The powers of water and of wind here revel in all their glory. (John Edwin Banks, June 12, 1849)

Massive in scale, each monument towers hundreds of feet above the floor of the North Platte Valley. Each is made of stacked layers of sand-stone, siltstone, mudstone, and volcanic ash. Differential erosion of these layers—some soft, some hard—gives each monolith a distinct shape. Courthouse Rock and Castle Rock are blocky and rectangular, reminis-cent of colossal buildings. Chimney Rock looks like an upside-down fun-nel. Its lower section consists of soft strata that erode into slopes, while the upper chimney section is composed of sandstone tough enough to stand as a vertical column.

The local Sioux knew what they saw in the towering phallus of Chim-ney Rock. They called it Elk Penis. This was too graphic for most white

4.4 Chimney Rock, in the valley of the North Platte River, western Nebraska. Chimney Rock is the most conspicuous geologic landmark on the emigrant trail and is mentioned in virtually every emigrant account. It stands 325 feet tall from base to tip.

sensibilities, even those of the rough fur traders who, in the 1830s, were among the earliest whites to report on the rock. They skirt around the native name in their accounts: "One of these cliffs is very peculiar in its appearance, and is known among . . . the natives as 'Elk Peak.'" "Arrived at the Chimney or Elk Brick, the Indian name." "We are now in sight of E.P., or Chimney Rock, a solitary shaft . . . one of the most notorious objects on our mountain march."

The emigrants were intensely curious about the rock monuments of the North Platte Valley. "The geological processes at work among these buttes are in fact of the highest interest," George Gibbs enthused. "How came such an immense pile so singularly situated?" Rufus Sage wondered at Chimney Rock. "What causes united their aid to throw up this lone column, so majestic in its solitude, to overlook the vast and unbroken plains that surround it?" "People say that it [Chimney Rock] was thrust up out of the ground, but no one agrees how," Rachel Larkin mused. Sage's and Larkin's comments show that some emigrants believed the monuments were somehow pushed up out of the Earth. In fact, they are the products of the opposite force—tearing down. They are landforms sculpted by the removal of rock. Many emigrants realized this. Chimney Rock, Edwin Bryant concluded, "is what remains of the bluffs of the Platte, the fierce storms of wind and rain which rage in this region having worn it into this shape." Joseph Stewart likewise realized that "the whole country about here appears to have been an extensive plain hundreds of feet above present levels, as if the soft marl and earthy limestone of which

4.5 A portion of ten-mile-long Scotts Bluff, in the valley of the North Platte River, western Nebraska. The flat-lying layers on the bluff face represent the eroded remains of the once-continuous sedimentary layer cake that formed the ancient Great Plains prior to the Exhumation. The wagon stands outside the Visitors Center at Scotts Bluff National Monument.

4.6 View southwest from the top of Scotts Bluff, looking out at the dissected land-scape of the North Platte Valley. The buttes in the distance are all that remains of the once-continuous sedimentary layer cake that stacked up on the Great Plains before the Exhumation.

it was composed had been washed away, leaving those remnants to show its former elevation."

Scotts Bluff marks the crescendo of the North Platte monuments, dwarfing the others in scale and magnificence. It stretches as an irregular 10-mile-long ridge across the south side of the valley, rising more than 600 feet above the braided course of the North Platte River. A road winds up to the top today, and the view takes in several thousand square miles of the western Great Plains. The Laramie Range—the beginning of the Rocky Mountains—looms 80 miles to the west. The view east skips across the tops of the other monuments and stretches back nearly to the Platte forks. The eagle's-eye perspective shows that the North Platte monuments are the carved-up remains of a thick layer cake of rock strata that once stretched unbroken for hundreds of miles across the western Great Plains. The North Platte River and its tributaries have now eaten most of this layer cake. The scraps they left behind form Scotts Bluff, Castle Rock, Chimney Rock, and other scattered fins, knobs, and spires. Restore the North Platte Valley to maximum cake, and all of the visible roads, rails, towns, and fields would be entombed under nearly 1,000 feet of layered sandstone, siltstone, mudstone, and volcanic ash.

THE SEDIMENTARY layer cake of the western Great Plains is made mostly of the ground-up debris of the Rocky Mountains. Ancient rivers swept this debris east out of the mountains and spread it as layers across the plains. The rivers kept up this work until about 5 million years ago. Then, suddenly, they changed their behavior. Rather than laying down more layers, the rivers began to bite down into the existing layers. The result is today's dissected landscape of towering rock monuments, ridges, and bluffs. To understand how and why this happened, we need to go back to the source—the uplift of the Rocky Mountains themselves.

The Foreland Ranges of the Rocky Mountains (the easternmost Rockies, including the Front Range, Laramie Range, Bighorn Range, and others) are giant blocks of basement rock pushed up along great faults during the Laramide Orogeny—a spectacular mountain-building episode that we'll explore in detail in chapter 7. The ranges rose in response to massive sideways squeezing of the Earth's crust. Just as a watermelon seed pops out when pinched between your thumb and forefinger, a mountain range can pop up (albeit incrementally and slowly) when the crust is squeezed from the sides. The ranges began rising about 75 million years ago and stopped rising about 30 million years later.

Erosion wears down mountains even as they go up. As the Foreland Ranges rose, erosion tore at their roofs and washed the broken debris down into the intermountain valleys, where it piled up as layers of sand and gravel. In some places more than 20,000 vertical feet of rock (nearly four miles) was torn from the rising ranges. You can see the results today throughout the Rockies as thick stacks of gravelly strata exposed in the valleys. Fragments of the youngest, highest rock of the mountains, the first to erode, are at the bottoms of the stacks. Pieces of the older, deeper rock of the mountains, later to erode, lie near the tops. The valleys hold the disassembled mountains in reverse order.

By about 38 million years ago, late in Eocene time, the intermountain valleys had filled high enough to let the eroded debris spill east across the Great Plains. The debris cascaded east in three major pulses, mapped today as the White River Group, Arikaree Group, and Ogallala Group.

Each debris pulse extended farther east as it piggybacked on the one below. Each consists of a thick stack of gravel, sand, silt, and mud laid down by braided rivers and windblown dunes. Meanwhile, far to the west, colossal volcanic eruptions periodically slung forth clouds of volcanic ash. Carried east by prevailing winds, the ash settled out of turbid skies like toxic snow, piling up thickly in depressions while spreading out more thinly over high spots. Today these ash beds stand out clearly as white horizontal stripes on the eroded bluffs, like thin layers of frosting between thicker layers of cake. (The magma source that spewed the ash now powers the geysers and hot springs of Yellowstone National Park—a story that we'll explore in chapter 9.)

Standing on the Great Plains in those days, you would have gazed across vast grasslands threaded by sluggish, sediment-choked braided rivers, flowing generally east. An exotic, Serengeti-like fauna of rhinoceroses, camels, archaic elephants, and small three-toed horses cropped the lush fare. They kept a wary eye out for doglike amphicyonids, primeval wolves, and saber-toothed cats. Death rained down periodically from ash-choked skies. At Ashfall Fossil Beds State Historical Park in western Nebraska, an entire herd of Miocene rhinoceroses lies entombed in 12-million-year-old volcanic ash.[3] Some of the rhinos still have grass seeds stuck between their teeth. Some rhino mothers still carry the remains of unborn young, and newborn rhino calves lie curled up next to adults (presumably their parents). The ash probably did not kill the animals outright. More likely, they died of starvation and thirst on the hellish post-eruption landscape. For thousands of square miles, the ash smothered grass, turned rivers to sludge, and filled the air with choking dust. Many animals probably died of slow lung failure. Volcanic ash contains high amounts of silica—like tiny shards of glass. Inhaled in large amounts, it can slowly lacerate lungs from the inside out, so that the animals may have drowned in their own lymph and blood.

By the end of Miocene time some 5 million years ago, the accumulating layers on the Great Plains had nearly buried the Rocky Mountains. The layers stacked up so thickly that they completely covered the lower parts of the ranges and reached for the chins of the higher peaks. Visualize a house with sand banked against one side almost to the roofline, and you

---

3. Volcanic ash, which is tiny bits of solidified magma (no relation to wood ash), contains traces of radioactive elements that change over time like a ticking clock. Radioactive uranium, for instance, changes into lead at a precise rate. When a volcano spews its ash over the landscape, the uranium-lead clock starts ticking. The more time goes by, the more of the original uranium shifts over to lead. By measuring the ratio of uranium to lead in a volcanic ash layer, we can figure out how long ago the ash erupted.

have an image of what the Laramie Range looked like 5 million years ago, with the sedimentary layers of the Great Plains banked up against its eastern flank. Five million years ago, you could have walked west across the Great Plains and stepped off directly onto Rocky Mountain summits.

That, of course, is not what the Rockies look like today. Now they rise from the Great Plains like a mighty wall. The reason is that rivers have recently stripped away the layers that once nearly covered the ranges. Starting about 5 million years ago, rivers began to slice down through the Great Plains layer cake, carving out spectacular monuments and bluffs along their eroding valleys, and uncovering the once-buried mountains in the process. Geologists call this momentous event the Exhumation of the Rocky Mountains, or simply the Exhumation. We'll visit more of its effects as we transit the mountains ahead. The Exhumation is the single most important landscape-shaping event in this part of the West—and it isn't over yet. The Exhumation is a work in progress, as every day the rivers bite still further into the land.

WHAT HAPPENED 5 million years ago to trigger the Exhumation? There are two competing theories.

One is regional uplift. This theory holds that beginning about 5 million years ago the Rocky Mountain region, along with much of the rest of the West, began slowly rising. As the land rose, the rivers responded to the steeper slopes by flowing faster. The invigorated rivers sliced down through the Great Plains layer cake, exhuming the Rockies and carving the landscape that we see today.

The alternative theory is climate change. This theory holds that about 5 million years ago the West grew wetter and cooler, coinciding with a general shift to cooler climates worldwide at about this time. More precipitation gave the rivers more erosive power, and so they began to cut downward. The result would be the same—the Exhumation of the Rockies and widespread dissection of the Great Plains layer cake.

Both theories make testable predictions. The uplift theory predicts that the Rocky Mountain region was several thousand feet lower before the Exhumation began. If the Rockies were not lower, then regional uplift cannot explain the Exhumation, leaving climate change as the more likely explanation. All we need to know is the elevation of the Rockies 5 million years ago, before the Exhumation began.

Unfortunately, there is no foolproof way to measure ancient elevations (called paleoelevations). One approach is to look at plant fossils, on the reasoning that different types of vegetation grow at different elevations. Research on fossil plants, particularly in the Florissant Fossil Beds of Colorado, suggests that the pre-Exhumation Rocky Mountains were at about the same elevation as today. In other words, if the plant data are correct, there has been no regional uplift of the Rockies, suggesting that the climate theory is the better explanation for the Exhumation.

But other more recent paleoelevation studies strongly support the uplift theory. One study looks at bubbles in lava flows. Bubbles expand under lower pressure, so lava that erupts at higher elevations should have larger bubbles. By recording ancient atmospheric pressures, lava beds can indirectly measure paleoelevation. Lava bubble data from the Colorado Plateau suggest that substantial regional uplift—more than 3,000 feet—has occurred since 5 million years ago. This is certainly enough uplift to explain the Exhumation, if comparable amounts of uplift also occurred in the Rockies.

Another study has tackled the paleoelevation question by measuring the slope of the Miocene Great Plains. Here's the idea: If the Rocky Mountain region has risen since the sedimentary layers that form the Great Plains were laid down, then those layers will today be tilting east at a greater slope than when they formed. Visualize a partly inflated balloon under one end of a long board. Pump air into the balloon (the rising Rockies) and the board (the Great Plains) rises on one end and tilts more steeply east. A large section of the pre-Exhumation Great Plains lies preserved in the large V between the North Platte and South Platte rivers. It's called the Cheyenne Tableland (see fig. 4.1), and it represents a section of the ancient Great Plains that has—so far—escaped the attack of rivers. The tableland rises west toward the Laramie Range at slopes that increase from about 10 feet per mile near the Platte forks to about 50 feet per mile near the mountain. How does this compare to the slope of the tableland before the Exhumation? It turns out that the size of gravel in braided streams is a fair measure of the slope of the streambed. Steeper streams move larger pieces of gravel. You can thus look at the size of gravel in ancient braided stream deposits to figure out the slopes down which those streams flowed. Streambed gravels from the Cheyenne Tableland show that pre-Exhumation streams flowed down slopes ranging from a half-foot to five feet per mile—much less than the slope of the tableland today. Conclusion: the tableland tilts east more steeply today than it did before the Exhumation. The difference translates into roughly 2,000 or more feet of uplift of the Rockies. This study then, like the lava bubble study, supports the uplift theory.

To sum up, both the lava bubble and the stream slope studies point to regional uplift as the main cause of the Exhumation. But this doesn't mean that climate change didn't contribute. Mountain glaciers first formed in Wyoming about 3 million years ago, showing that the mountain peaks had begun to hold more precipitation. We don't know whether these glaciers formed because the peaks were uplifted to higher elevations or because climate change increased winter snowfall. Either way, more snowmelt would have given the rivers more power to carry out the Exhumation.

IF YOU FOLLOW the historic overland trails west, you'll find that you are within shouting distance of a highway or a railroad track much of the time. This makes sense. Early explorers found the smoothest routes over rough country, and so today's roads and rails often follow yesterday's trails. Interstate highways follow various emigrant trails for hundreds of miles across Nebraska, Utah, Idaho, and Nevada. But not in Wyoming. Across most of Wyoming, Interstate 80 tracks far south of the Oregon-California Trail, cutting directly across the center of the Laramie Range rather than following the emigrant trail on its long arc around the north end of the range. Why?

The reason lies in the Exhumation—or more precisely, in the unfinished work of the Exhumation. To cross the Laramie Range, Interstate 80 and the railroad take advantage of the Cheyenne Tableland—that large

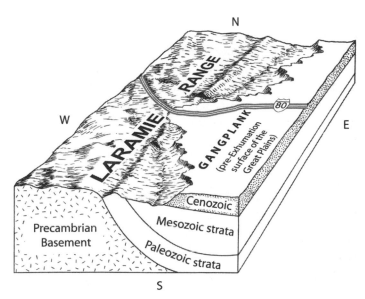

4.7 Interstate 80 and the transcontinental railroad cross the Laramie Range by taking advantage of the Gangplank—a narrow strip of the pre-Exhumation (Miocene) Great Plains that has so far escaped river erosion. The Gangplank surface takes the interstate and the rail lines smoothly from the Great Plains up to the crest of the range. The Oregon-California Trail took a longer route, following the North Platte River around the north end of the range. Reproduced by permission of the Wyoming State Geological Survey.

section of the pre-Exhumation Great Plains that still exists between the North Platte and South Platte rivers. The tableland acts like a great plank, with one end on the low eastern plains of Wyoming and the other end on the roofline of the Laramie Range. When surveyors for the Union Pacific Railroad in the 1860s found this natural ramp over the Laramie Range, they were delighted. Forget laying the track along the old emigrant trail, they said—it's too far. They laid the rails on the rising tableland surface and dubbed it the Gangplank.

The Cheyenne Tableland/Gangplank offers the smoothest possible crossing of the Laramie Range. It is also a more direct route west than the long, curving route of the North Platte around the north end of the range. Why didn't the emigrants take advantage of this shortcut? The answer is that they were beholden to river valleys—the keepers of the trinity (grass, wood, and water). The Cheyenne Tableland/Gangplank has no large rivers. Indeed, it exists precisely *because* there are no large rivers chewing it up and sending it in tiny pieces to the Gulf of Mexico. The North Platte River was a known entity—a reliable source of water and grass that led toward South Pass. It came at a price, though. It added distance to the journey, and it doomed the emigrants to the horrendous pull through the Black Hills (next chapter). In 1857, as gold rush emigration was waning, explorers blazed a viable wagon route across the Cheyenne Tableland. The route followed Lodgepole Creek up to the central crest of the Laramie Range, thereby bypassing the North Platte route entirely. But the Lodgepole Trail was discovered too late to see much use. Had it been scouted a decade earlier, many emigrants would probably have favored this smoother, shorter road west, where a piece of the ancient Great Plains still lies uneaten by the hungry rivers.

* 5 *

BLACK HILLS
AND BENT ROCK

Although going overland to California was much cheaper than going by ship, it was still a costly undertaking. An emigrant had to buy all of the necessary equipment and supplies, bring enough money or bartering goods to cover expenses along the way (replacement livestock, ferry tolls, Indian trading, and other incidentals), and have enough left over to get started at the other end. Emigration usually meant losing farm or other income for at least one and usually several years. Yet even before James Marshall saw that telltale golden glimmer in his millrace in 1848, emigrants were heading overland in increasing numbers to both Oregon and California. Many left behind lives of reasonable comfort, forsaking extended communities of kin and friends. During the 1840s Horace Greeley, the eminent editor of the *New York Daily Tribune*, regularly poured out his scorn for these restive souls.

> For what, then, do they brave the desert, the wilderness, the savage, the snowy precipices of the Rocky Mountains, the weary summer march, the storm-drenched bivouac, and the gnawings of famine? Only to fulfill their destiny! There is probably not one among them whose outward circumstances will be improved by this perilous journey.

This was the same Greeley who had advised young men to "go West and grow up with the country." But Greeley's west was the Midwest—the fertile lands of Ohio, Indiana, and Illinois. Greeley saw little value or purpose in transcontinental emigration. No less curmudgeonly than Greeley were the editors of the *Missouri Republican* in 1844.

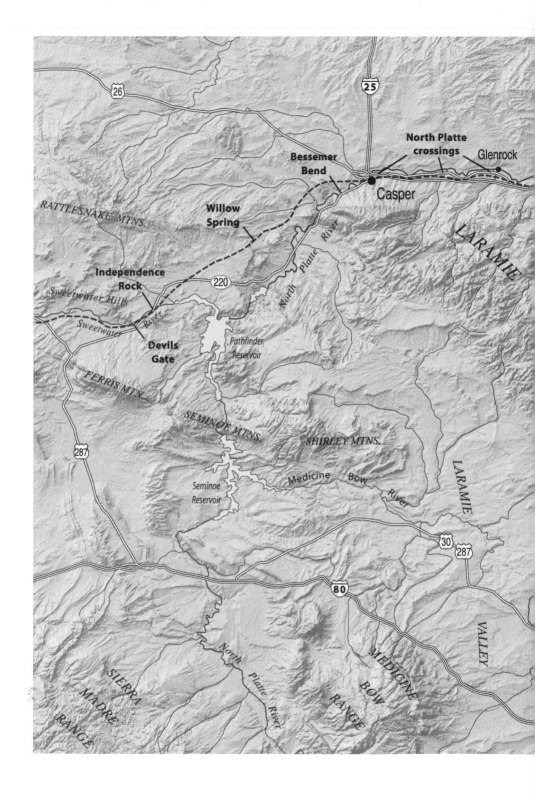

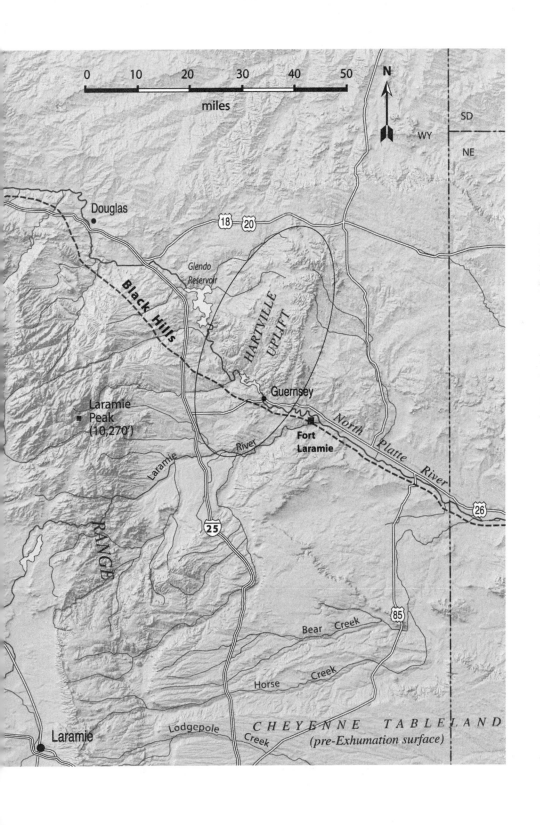

No man of information or in his right mind, would think of leaving such a country as this [Missouri], to wander over a thousand miles of desert and five hundred of mountains to reach such as that [Oregon].

What made people head west in rising numbers, even before the days of gold? One important factor was disease. Midwestern farmers, especially those in the swampy bottomlands along the Missouri and Mississippi rivers, regularly endured outbreaks of malaria, smallpox, flu, and cholera during the 1830s and 1840s. Western promoters described Oregon and California as disease-free paradises.

Another push was the financial panic of 1837, whose effects spilled forward into the 1840s. By the 1830s, midwestern farmers were producing huge amounts of food—far more than the economy could absorb once it entered the 1837 depression. Prices for farm goods plummeted. Wheat sold for far less than the cost of raising it; prices of bacon and lard fell so low that river steamboats burned it for fuel. Meanwhile the U.S. population was surging, with farms pushing out to the Missouri River. By the 1840s, the combination of economic pressure and land hunger was pulling more and more people overland to Oregon or California.

And the land was there for the taking—as long as one ignored Indian claims to it. The Preemption Act of 1841 held that anyone who squatted on public land for fourteen months had first right, once the land was surveyed, to buy up to 160 acres at a set minimum price. The Preemption Act presaged the Oregon Donation Land Claim Act of 1850 and the famous Homestead Act of 1862. These laws granted free land, up to certain limits, to anyone who demonstrated commitment to living on the land and improving it. Even during the gold rush years, many emigrants headed west for land, not gold. California promoters stoked the emigration fires with hyperbolic descriptions of the mild climates, fertile soils, and the beauty of the Sacramento Valley. Lansford Hastings, for instance, claimed that compared to California, "the deep, rich, alluvial soil of the Nile, in Egypt, does not afford a parallel."

Beyond these concrete motives for emigration, there was the simple allure of new beginnings in an expansionist age—an age of Manifest Destiny. It is "the fulfillment of our manifest destiny to overspread the

5.1 (*previous pages*) West of Fort Laramie, the emigrant trail enters rougher terrain as it follows the North Platte River upstream around the north end of the Laramie Range. The river cuts several impassable canyons through a ridge called the Hartville Uplift. The canyons forced the emigrants onto a rough detour through the Black Hills (the Laramie Range foothills). The trail then leaves the North Platte River near present Casper, Wyoming, and crosses overland to the Sweetwater River. The roughness of the landscape, in which rivers punch through ridges and uplifts rather than going around them, reflects the burial and exhumation history of the Rocky Mountains.

continent allotted by Providence," journalist John O'Sullivan wrote in 1845, thus coining the famous phrase. Many saw westward expansion as America's God-given right, and to hell with those who claimed other-wise—like Mexico. "What has miserable, inefficient Mexico—with her superstition, her burlesque upon freedom, her actual tyranny by the few over the many—what has she to do with the peopling of the new world with a noble race?" Walt Whitman sneered in 1846. At the end of the Mexican War in 1848, Mexico ceded to the United States most of what we today call the Southwest—the territory from Texas to California—fur-ther feeding expansionist fever.

The age of Manifest Destiny was manifestly restless. "An American will build a house in which to pass his old age and sell it before the roof is on," the French historian Alexis de Tocqueville remarked in the 1830s. "He will plant a garden and rent it just as the trees are coming into bearing; he will clear a field and leave others to reap the harvest. . . ." Emigrant James Clyman saw this restive spirit in his fellow pioneers. Returning east along the overland trails in 1846, Clyman passed hun-dreds of men and women heading toward Oregon and California, and he wrote, "It is remarkable how anxious these people are to hear from the Pacific country and strange that so many of all kinds and classes of People should sell out comfortable homes in Missouri and Elsewhere pack up and start across such an immense, barren waste to settle in some new place of which they have at most so uncertain information, but this is the character of my countrymen."

LOOKING WEST from Scotts Bluff in the valley of the North Platte River, the emigrants got their first good look at the Laramie Range—the begin-ning of the Rocky Mountains. "We are now in sight of the highest por-tion of earth that I ever looked upon," William Cornell wrote as he gazed at 10,270-foot Laramie Peak. "Its snow covered top formed a cheering contrast to the monotony that had marked our view for so long," Sarah Royce remembered. From this point forward, the emigrants would never be out of sight of mountains.

The Laramie Range, like the rest of the Rocky Mountain Foreland Ranges, is a block of deep, ancient basement rock squeezed upward through younger overlying layers of Paleozoic and Mesozoic strata. To-day these layers—once flat-lying, but now shouldered out of the way—lean like tilted books against the mountain. To see this bent and broken rock, tossed aside by the rising core of the mountain, is to witness the stupendous forces that built the Rockies.

The western Great Plains roughen as they rise toward the Rockies. Aridity asserts itself more boldly on the land, and grass gives way inexo-rably to sagebrush and greasewood. William Swain noticed the countryside "becoming very hilly; the streams rapid, more clear, and assuming the character of mountain streams. The air is very dry and clear, and our path is lined with wild sage." Edwin Bryant didn't much like the change: "The general aspect of the scenery is that of aridity and desolation. The face of the country presents here those features and characteristics which proclaim it to be uninhabitable by civilized man."

Whatever Bryant felt about its habitability, the western Great Plains were most certainly inhabited. The forks of the Platte corresponded roughly to a transition point in the distribution of Great Plains Indians. East of the forks, the emigrants had mostly encountered the village farming communities of the Pawnee. West of the forks they met with the Sioux and Cheyenne, horse-riding nomads pursuing buffalo herds. Most emigrants had a general disdain for Indians, compounded by mistrust associated with the threat (usually highly exaggerated) of Indian attack or robbery. Nonetheless, many could not help but be impressed by the good-looking Sioux they met on the high western plains.

> The pretty young squaws certainly are beauties of Nature and will compare very well with some of the young ladies of the States who think themselves handsomer. . . . They are what nature designed them to be, women without stays or padding. (Andrew Griffith, June 3, 1850)

> Many of these women, for regularity of features and symmetry of figure, would bear off the palm of beauty from some of our most celebrated belles. . . . Their complexion is a light copper color, and . . . the natural glow of the blood is displayed upon their cheeks in a delicate flush, rendering their expression of countenance highly fascinating. . . . The men are powerfully made and possess a masculine beauty which I have never seen excelled. (Edwin Bryant, June 23, 1846)

ARGUABLY THE most transforming event in the history of the Sioux — before the arrival of white emigrants — was the introduction of the horse. Horses first arrived in the West with Coronado's 1540 expedition, and within two centuries they had spread widely across the Great Plains. The animal revolutionized Indian hunting and warfare, and led directly to a mobile, nomadic existence. In effect, through horses, Plains Indians harnessed the power of grass — a source of energy limited only by the number of horses you could acquire and manage. The tribes rode that power to regional dominance — but not for long. Grass power led directly to their downfall by fueling the westward expansion of white emigrants.

Wild West mythology often portrays Indian-emigrant relations as violent and hostile. But the celluloid image of circled wagons surrounded by whooping Indians does not ring true with most emigrant accounts.

(Emigrants did commonly circle their wagons at night—to corral their livestock, not for defense. "When we camp at night, we form a corral with our wagons and pitch our tents on the outside, and inside of this corral we drive our cattle," Sallie Hester explained.) Prior to 1849, most emigrant accounts describe interactions with Indians as friendly. The natives gave directions and tips about upcoming terrain and sources of water and grass, helped on treacherous river crossings, and engaged in mutually beneficial trading. In the early years of the westward migration, it seems that native peoples did not see the small number of transient emigrants as a threat.

The gold rush changed that. Gold fever unleashed a white tsunami across native lands, eroding emigrant-Indian relations and muddying heretofore friendly waters. Hordes of emigrant cattle stripped grass from river valleys. Emigrant campfires burned up sparse wood supplies. The noise and chaos of the wagon trains probably drove game away from the river valleys. Disease followed the wagons and stayed when they had gone. Unhappy Indians began to demand payment for passage across their lands. Most emigrants scoffed at the notion that Indians could lay claim to land. Armed to the teeth and banded into large trains akin to mobile armies, many emigrants refused to pay such tribute and threatened reprisal if Indians pressed too hard. Confrontations and general hostility escalated. Even in the worst years, though, far fewer emigrants died from Indian attacks than died from disease. In the 20-year period from 1840 to 1860, there are 362 documented instances of emigrant death from Indians. Estimates of total emigrant deaths during the same period range from 10,000 to 30,000. In other words, Indians probably caused somewhere from 1 to 4 percent of emigrant fatalities. More Indians died at emigrant hands than vice versa; in that same 20-year period, there are 426 documented reports of Indians killed by emigrants.[1]

"YOU ARE NOW 640 miles from Independence, and it is discouraging to tell you that you have not yet traveled one-third of the long road to Oregon." So the 1846 *Shively Guide* informed emigrants of their progress upon arriving at Fort Laramie, on the high western Great Plains under the sunset shadow of the Laramie Range.

Fort Laramie lies at the confluence of the Laramie and North Platte rivers. Established as a trading post by trapper William Sublette in 1834, the fort was bought by the U.S. government in 1849 and expanded into a military post. As relations between whites and Indians worsened through the 1850s and 1860s, Fort Laramie became a key staging point for the

---

1. Data from Unruh 1979, table 4, p. 185.

U.S. Army's anti-Indian campaigns. Before that unhappy time, though, the fort was a bustling center of commerce on the western Great Plains. Crowds of Sioux and white traders lived in sprawling nomadic camps around the fort, bartering and bargaining with each other and with passing emigrants.

After six weeks of prairie wilderness, Fort Laramie was a sight for sore emigrant eyes. Here emigrants often camped for several days, swapping stories, writing letters home, and buying what supplies they could afford. "Oh, what a treat it does seem to see buildings again," Lucy Cooke enthused.

> My dear husband has just been over to the store there to see if he could get anything to benefit me, and bless him, he returned loaded with good things, for which he had to pay exorbitantly. He bought two bottles of lemon syrup for $1.25 each, a can of preserved quinces, chocolate, a box of seidlitz powders, a big packet of nice candy sticks, just the thing for me to keep in my mouth, and several other goodies. . . . The preserved quinces seemed so grateful to my poor throat, and I took such tiny swallows of each of them, and then hung the can up overhead to the wagon bows.

West of Fort Laramie, the land rises rough and calloused toward the foothills of the Laramie Range. Ledgy bluffs of Arikaree Group strata— leftover bits of the Great Plains layer cake—pop up like stone scabs across the landscape. "With the change in the geological formation on leaving Fort Laramie, the whole face of the country has entirely altered its appearance," John Van Tramp observed.

> Eastward of that meridian, the principle objects which strike the eye of a traveler are the absence of timber, and the immense expanse of prairie, covered with the verdure of rich grasses, and highly adapted for pasturage. Whenever they are not disturbed by the vicinity of man, large herds of buffalo give animation to this country. Westward of Laramie River, the region is sandy, and apparently sterile; and the place of the grass is usurped by the artemesia, and other odiferous plants, to whose growth the sandy soil and dry air of this elevated region seem highly favorable.

In the evermore-dissected terrain, emigrants had to veer away from the North Platte River for long stretches, with tough pulling over hills and across ravines. One mile south of present Guernsey, Wyoming, an outer bend of the river runs up against an 80-foot-high bluff of Arikaree sandstone. Unable to pass along the riverbank, everyone had to climb up and over the bluff. All wagons followed the same narrow track over the bluff, excavating a five-foot-deep defile today called Deep Rut Hill. A few miles away, on the inside of a wide bend in the North Platte, stands

5.2 Deep Rut Hill, near Guernsey, Wyoming—probably the oldest road-cut west of the Mississippi. Thousands of wagon wheels and hooves gouged these ruts into the soft sandstone as they passed over a bluff along the North Platte River.

Register Cliff—a long wall of Arikaree sandstone riddled with emigrant signatures. The soft rock is easily gouged with a stick or knife, and emigrants often carved their names with great calligraphic flare. *G.O. Willard, Boston, 1855. H.H. Roman, 1852. C.W. Bryson, 1855, KY. S.H. Patrick, June 6 1850.* The names go on for more than a hundred yards along the cliff face. Other names are here too, far more numerous though not so worthy. They often cut across and obscure original emigrant signatures. *Bubba rules. Chris and Tracy 2001. Missy + Mona Peters June 1996. Dennis + Carolyn Luescher, Reno NV 1984.*

West of Register Cliff and Deep Rut Hill, further travel along the North Platte becomes impossible. For many miles upstream from present Guernsey, the river winds mostly in the depths of inaccessible canyons framed by sheer walls of red sandstone and limestone. To bypass the canyons, the emigrants had to veer off along a detour that would take them face-to-face with the elephant.

For the previous 700 miles, the emigrants had rolled west with comparative ease along the gently rising valleys of the Platte and North Platte rivers. The rivers provided water, the grassy bottomlands provided feed for livestock, and the smooth slopes of the river valleys formed natural avenues for overland travel. Now, swinging away from the North Platte, they entered the rough foothills of the Laramie Range — a region called the Black Hills.[2] Countless ravines, carved out by streams charging down from Laramie Peak, cut directly across the trail. It was like crossing a landscape clawed by a gigantic bear, and according to Sallie Hester, it made for "sixty miles over the worst road in the world." Many emigrants abandoned their wagons in the steep ravines and pressed on with whatever they could pack on the backs of their animals. "Every day we pass good wagons that have been left for any one that might want them," Margaret Frink observed. A. J. McCall found the road "strewed with provisions which have been cast out to lighten loads," and noted bitterly, "One vagabond who had an oversupply of sugar and coffee, instead of placing them where they might do somebody good, scattered these desirable articles, as he went, in the dust. Such a wretch richly deserves starvation." McCall cataloged a stunning assortment of abandoned goods in the Black Hills. "Bar iron, black-smiths' anvils, bellows, crowbars, drills, augers, gold washers, axes, lead, trunks, spades, baking ovens, cooking stoves, kegs, harness, clothing, bacon and bread. Every day one could select from the debris a complete outfit."

For those who persevered, it took about six days to muscle the wagons through the up-and-down landscape before weary eyes finally spied the North Platte River, winding peacefully across flat bottomlands upstream from where it enters the canyons. "Mighty glad were we to once again behold the good stream which had so long furnished us good roads and fine grass, and so many times quenched our thirst with its healthy waters," William Swain wrote after dipping his cup once again into the river.

THE TRIAL of the Black Hills didn't stop some emigrants from pondering the spectacular displays of bent and riven rock all around. "How I wish I was a geologist," James Berry Brown mused, "then these rambles over rocks and hills would be of some benefit to me." A strange piece

---

2. The name (no relation to South Dakota's Black Hills) comes from the dark stands of pine, spruce, and cedar that dot the slopes. Travelers on both the north and south sides of the North Platte had to swing away from the river to bypass the impassable canyons. South-side travelers had to pass through the Black Hills. North-side travelers after 1850 could stay on the north side of the river along a route called Child's Cutoff. Either way was rough.

of geologic happenstance forced the emigrants into the Black Hills. The detour through the hills comes where the North Platte River crosses the Hartville Uplift—a broad hump of arching rock strata, 30 miles across, rising like an oval blister on the land. The river, rather than taking an open route *around* the uplift to the north, slices straight *through* it, cutting three deep canyons in quick succession between today's towns of Orin and Guernsey, Wyoming. Had the river gone around, there would have been no canyons, and the emigrants could have just rolled west along the bottomlands as before.

Why would the North Platte cut sheer canyons through the Hartville Uplift when it could have gone around? The river's choice makes about as much sense as sledgehammering your way through a stone wall instead of walking around it. Yet the North Platte is not alone in this headstrong behavior. Many Wyoming rivers punch straight through ridges and up-lifts—even where an easy way around lies nearby.

The Wind River in central Wyoming flows calmly across the Wind River Basin before barreling headlong through the Owl Creek Mountains, cutting 2,000-foot-deep Wind River Canyon. Its behavior is so strange that early explorers did not even recognize it as the same river where it emerges on the other side and named it there the Bighorn River. The Laramie River runs north down its valley on the west side of the Laramie Range, seemingly on track to join the North Platte River. Then suddenly, as if on a whim, the river turns east and slashes, foaming and white, straight through the center of the Laramie Range. Not to be outdone, the North Platte punches through no less than *three* major obstacles on its way to the Great Plains. First it cuts through the Seminoe Mountains to make Seminoe Canyon, then through the eastern Granite Mountains to make Fremont Canyon, and then through the Hartville Uplift to form Glendo, Wendover, and Guernsey canyons. In fact, you would be hard-pressed to find a river *anywhere* in the central Rockies that does not cut through a large rock obstacle somewhere along its route. In Wyoming alone, the Sweetwater, Snake, Shoshone, Bighorn, and Belle Fourche join the list of rivers that slice through rock ridges rather than going around.

How can thin ribbons of water charge barriers of solid rock and come out victorious on the other side? The answer comes into focus only when viewed through the lens of time. Geologists think Wyoming's rivers have been able to cut through mountain ridges because they once flowed high *above* these obstacles. Recall from the last chapter that after the uplift of the Rockies, the mountains underwent a long period of erosion and burial. By 5 million years ago, the ranges were buried up to their eye-brows in their own sedimentary debris. The Wyoming landscape then looked much like Iowa today, with a few low peaks added. The broad

5.3 Two examples of canyons cut by stream superposition. The upper image shows the Bighorn River cutting through Sheep Mountain in northern Wyoming. The lower image shows the North Platte River cutting through the Hartville Uplift in southeastern Wyoming. The view in the lower image looks downstream along Guernsey Canyon from the top of a dam that forms Guernsey Reservoir. This canyon, and two more upstream, forced the emigrants onto a 60-mile detour through the Black Hills—the foothills of the Laramie Range. Upper photo courtesy of Peter Huntoon.

Wyoming about 5 million years ago
(end of Miocene time)

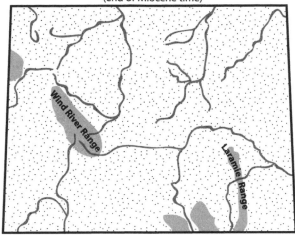

Wyoming today

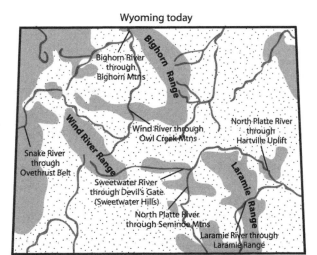

5.4 The upper figure shows what Wyoming probably looked like about 5 million years ago, at the maximum level of valley fill prior to the Exhumation. Only the highest mountain summits poked above the layers of sand and gravel. Rivers established paths across this buried landscape, often by chance crossing over hidden mountains. During the Exhumation, the eroding rivers cut down through the once-buried mountains to establish the paths that we see today. Note the locations of the Hartville Uplift and of Devils Gate, discussed in the text.

gravel plains gave no hint of smaller summits and ridges buried far below. Lazy rivers wandered eastward across these plains, oblivious to the buried ridges and often by chance crossing right over them.

Then came the Exhumation. The land rose, and the rivers cut downward into the stacks of sand and gravel. Where the rivers happened to

5.5 Devils Gate, the most striking example of stream superposition on the emigrant trail. Here the Sweetwater River has cut a 300-foot-deep chasm through a granite ridge, even though a clear route around the ridge lies less than one-half mile to the south (*right*).

cross above buried ridges, they worked their way down to the ridgetops and kept going, slicing through like chain saws to establish the paths that we see today. The pattern of today's rivers only makes sense when we recognize that they inherited their paths from ancestral streams that flowed high *above* the ridges, in what is now the blue Wyoming sky. In 1875 the great western explorer and geologist John Wesley Powell dubbed this process stream superposition.

Confronted by impassable canyons hacked by superposed rivers through once-buried ridges, the emigrants had no choice but to find a way around. The Black Hills detour is one example. Another canyon barrier would loom up 160 miles farther west, at Devils Gate on the Sweetwater River. Here the Sweetwater cuts a 300-foot-deep chasm through a granite ridge, even though an easy way around lies less than one-half mile away. A. J. McCall was one of many emigrants who pondered the river's decision. "It is difficult to account for the river having forced its passage through rocks at this point when a few rods south is an open level plain over which the road passes."

Today we understand that the Sweetwater blithely ignores topography because its ancestor flowed on a now-vanished landscape high above the

ridge. The emigrants could not know this, of course. No matter—they had their own ideas about Devils Gate. McCall proposed that the ridge "had been rent by an earthquake." Lucy Cooke thought that it had been "riven in two by some great convulsion of nature." Dr. Charles Parke suggested that the river had indeed gone around the ridge once, but then "there was fire below, but when the fire went out the crust cooled so rapidly, this mountain rib contracted and cracked in twain, allowing the stream a shorter cut to the valley below." Alonzo Delano proposed that the chasm had been wrenched open "by volcanic force," and some other emigrants agreed. Others favored water, believing that the river had somehow busted its way headlong through the ridge, perhaps by forcing its way through a fissure. Still others were confident that such a marvel had an altogether different explanation. "This is indeed wonderful to look at," Martha Missouri Moore wrote of Devils Gate, "and one stands in awe of Him Who tore asunder the mountains and holds the winds in the hollow of His hands."

ONE MOMENTOUS geologic event—the Exhumation—ties together much that we have seen up to now on the road west. The river dissection of the Great Plains; the carving of the spectacular rock monuments of the North Platte Valley; the rebirth of the Rockies as high mountains after their near-complete burial; and the cutting of numerous canyons through

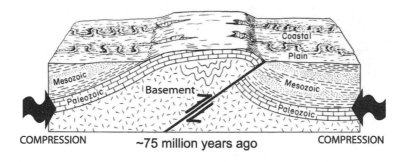

COMPRESSION   ~75 million years ago   COMPRESSION

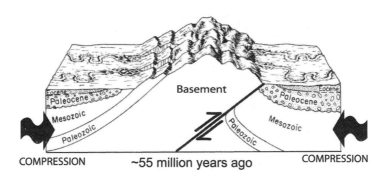

COMPRESSION   ~55 million years ago   COMPRESSION

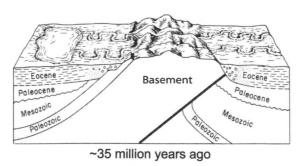

~35 million years ago

5.6 Today's range-crossing rivers and subsummit surface both owe their existence to the Exhumation of the Rocky Mountains. The Foreland Ranges began to rise about 75 million years ago in response to sideways compression. Rock debris eroding from the mountains filled up the valleys and continued to do so after the ranges stopped rising. By about 5 million years ago, the debris had nearly buried many of the mountains. Rivers established paths across the buried mountains, and pediments were notched around some of the exposed peaks. With the onset of regional uplift about 5 million years ago, the rivers sliced through the once-buried mountains, and the pediments were raised to their current elevation (approximately 10,000 feet) to become today's subsummit surface. Reproduced by permission of the Wyoming State Geological Survey.

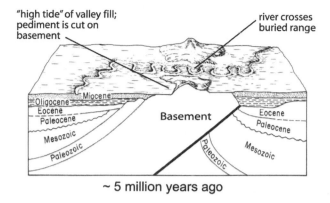

"high tide" of valley fill; pediment is cut on basement

river crosses buried range

Oligocene · Miocene
Eocene
Paleocene
Mesozoic
Paleozoic

Basement

Eocene
Paleocene
Mesozoic
Paleozoic

~ 5 million years ago

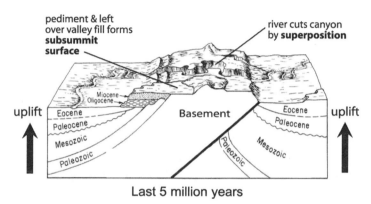

pediment & left over valley fill forms **subsummit surface**

river cuts canyon by **superposition**

uplift

Miocene
Oligocene
Eocene
Paleocene
Mesozoic
Paleozoic

Basement

Eocene
Paleocene
Mesozoic
Paleozoic

uplift

Last 5 million years

ridges by stream superposition—all of these come back to the Exhumation. Extraordinary events demand extraordinary evidence. For that, we leave the rivers and climb the mountains.

As a rule, mountain slopes get steeper going up. The lower flanks rise gently while the upper slopes are more precipitous. But many Rocky Mountain ranges don't look like this. Instead, at about 10,000 to 11,000 feet elevation, the steep slopes often flatten out at an eroded plateau, like a step cut below the high peaks. This landmark, called the subsummit surface, lies several miles wide and dips gently outward. Erosion has taken much of it away, but where it remains, your eye can trace it from range to range across intervening valleys like a high-tide line on islands in a bay.

The subsummit surface is indeed a high-tide line, but of rock, not water. It records the maximum burial level of the Rockies before the Exhumation. Stand on a Wyoming mountain and refill the valleys before you with sand and gravel up to the level of the subsummit surface. You look then onto a Miocene world—a time of three-toed horses and primeval rhinos; a time when the higher peaks poked like islands above the gravel sea and the lower peaks stood like seamounts below its surface.

The clearest exposures of the subsummit surface in Wyoming lie on the east face of the Medicine Bow Range and the southwest face of the Wind River Range. On one of my first trips to Wyoming, I tore off eagerly to the Medicine Bows to see it—finally—*the subsummit surface! The evidence that the mighty Rockies were once buried up to their chins!* I switchbacked up steep roads to 10,000 feet, jumped out, camera at the ready . . . and stared at bland forests of rangy pines. Wrong approach, I soon realized. If a flea wants to see the dog, he has to jump off. Back down the mountain I went, veering through the switchbacks, driving east across the 50-mile-wide Laramie Valley and up to the crest of the Laramie Range. The Medicine Bow Range lay to the west, sprawled end to end across the horizon. It rose up from the dusty valley, its flanks shifting from tan to dark green as pines took over from grass and sagebrush. Up it went, green and steep—and then it stopped flat, as if cut with scythe, at a forested plateau 10,000 feet high and more than 10 miles wide. The subsummit surface—exactly where I had stood several hours before. It was plain as day now, viewed from 50 miles away. Beyond it, 2,000 more feet of mountain climbed skyward, all bare rock and bright snow. It rested above the subsummit surface like a model mountain set on a tabletop, still looking like an island in a pre-Exhumation gravel sea.

In the view of many Rocky Mountain geologists, the subsummit surface is a relict pediment—an eroded bench cut by streams onto bedrock at the foot of a retreating mountain front. At the maximum level of valley fill prior to the Exhumation, streams apparently notched pediments around many exposed Rocky Mountain peaks. Then came the Exhumation. The streams stripped away the loose layers of sand and gravel that filled the valleys, leaving behind the hard bedrock pediments to become today's subsummit surface (fig. 5.6). While slower to erode than the valley fill, the subsummit surfaces have nonetheless been thoroughly cleaved up by stream canyons and Ice Age glaciers. A few more million years of erosion and they will probably disappear altogether, joining the endless procession of ground-up riverbed rock rolling out of Wyoming.

It has taken about 5 million years for the Exhumation to carve Wyoming's present landscape—5 million years to bulldoze out several thousand feet of valley fill and transform an Iowa-like alluvial plain into the rough, dissected landscape that we see today. What is 5 million years? From a human perspective, it's a mighty a long time. Five million years is 50,000 centuries—about 1,000 times older than the Giza pyramids, 10,000 times older than Columbus's arrival in the Americas. Five million years ago, there were no people yet on Earth. Back then, in a corner of Africa, the evolutionary lineage that would eventually lead to humans had just split from the one that would produce the great apes. The entire

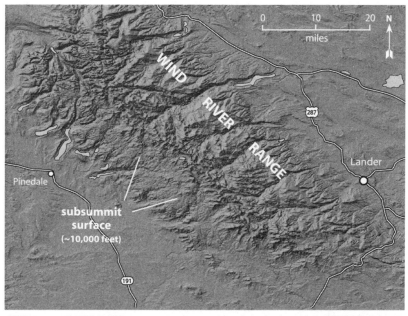

5.7 The subsummit surface on the southwest flank of the Wind River Range. The surface forms a broad plateau at about 10,000 feet elevation, like a bench cut into the side of the range. Glaciers that grew on the range during the ice ages carved deep valleys into the once-continuous surface. The photograph shows a portion of the surface southwest of the Green River Lakes. Courtesy of David Lageson.

span of human evolution took place during the Exhumation. Yet if you could touch the ends of geologic time with your outstretched arms, your left fingertips at the Earth's conception 4.55 billion years ago and your right fingertips at today, the time since the Exhumation began would take up 1/12th of an inch—a fingernail about ready for trimming.[3]

*

After passing through the gauntlet of the Black Hills, the emigrants gained the North Platte River again in the area of Deer Creek (near present Glenrock, Wyoming). They were now on the north side of the Laramie Range, and their route upstream would take them west past the mountain. Somewhere in the next 40 miles—between Deer Creek and Bessemer Bend (about 18 miles west of present Casper)—they had to cross to the north side of the river. Further travel upriver beyond Bessemer Bend would take them south, whereas the trail lay west, overland toward the Sweetwater River.

The crossing of the North Platte—if one did not take a ferry—was a frightful affair. "Men are daily drowned," Franklin Langworthy reported in June 1850. "Not one in a thousand can save his life by swimming, no matter how expert a swimmer. The water is cold, being formed from the melting snows, and the current rolls, boils, and rushes along with a tremendous velocity." Here, some 300 miles upstream of the Platte forks, the North Platte is no longer the shallow, braided, half-mile-wide stream of the Great Plains. It sweeps around the north end of the Laramie Range as a single deep, swift channel several hundred feet wide. Prior to 1847, the only way to cross the North Platte was by fording it, and emigrants in those early years fished many a drowned comrade from the swift waters. In 1847, as Mormons settled the Salt Lake Valley, Mormon entrepreneurs recognized a business opportunity on the North Platte. They established several ferries in what is now the Casper area, charging three to five dollars per wagon. By 1849 not even multiple ferries could keep up with the arriving wagons, and the trains backed up for miles, sometimes waiting for days to cross. Some emigrants refused to wait, as Margaret Frink observed in 1850:

> Some are making ferry boats of their wagon bodies taken off the wheels, and launched in the water, with long ropes to haul them back and forth across the river. In some cases, empty casks are tied to the four corners of the wagon body, to keep it from sinking. This plan is very dangerous

3. This metaphor for the Earth's history fitted to the span of outstretched arms comes from McPhee 1998, 69.

5.8 Ferrying across the North Platte at Deer Creek (east of present Casper, Wyoming), as portrayed by J. Goldsborough Bruff on July 20, 1849. Reproduced by permission of the Huntington Library, San Marino, California.

in the swift current, and we hear of many persons who have lost their lives in these attempts.

In 1851 Jean Baptiste Richard, a French mountain man, built a log toll bridge over the North Platte. In the spring of 1852, the snowmelt-charged river slapped it aside. Richard came back and built a better bridge, and soon put the Mormon ferries out of business. Richard and his partners had no need to go to California. They found gold in thousands of emigrant pockets right on the banks of the North Platte.

Once across the river, the emigrants faced 50 desiccated miles of sagebrush plains as they passed overland to the Sweetwater River. It was a bleak crossing—as Henry Bloom described:

> No sign of grass today whatever; a perfectly desolate and barren region. . . . Nothing grows here but the wild sage. Saw dead horses and oxen, lots of them today. I have seen in the last few days lots and lots of homesick chaps, many of them nearly discouraged; a fretful time in which men begin to show their real character; a discouraging prospect truly.

A new phenomenon, born of the growing aridity, greeted the emigrants here—alkali lakes. These shallow pools hold water briefly after rainstorms but otherwise are bone dry. They are coated with white crusty residues of saleratus and salt. Saleratus is potassium or sodium bicarbonate—essentially impure baking soda—and many emigrants "collected this deposit and used it . . . for the purpose of making bread light and spongy." But the alkali lakes were deadly for livestock. "We do not let our cattle taste it [the lake water], fearing its consequences," J. R. Starr explained, adding, "We saw a great many cattle lying dead and a great many that had been left which were yet alive." Oxen were particularly vulnerable to the alkali water. According to Oliver Goldsmith, it "seems to eat the lining of their stomachs. They lie down and soon commence bleeding at the nose, in which state there is no hope for them." Festering, maggot-riddled oxen carcasses did not improve travel along this stretch of the trail. Niles Searls believed a "blind man might find his way by the odor of the oxen."

On the high, dry Wyoming plains, the Earth sheds her former grassland modesty and bares her rocky skin, wrinkled by time and mountain upheaval. The beveled edges of bent, tilted strata poke up everywhere as irregular fins and ridges. At Avenue of Rocks, about 20 miles west of present Casper, the trail twists between ghoulish hogbacks cut on up-ended sandstone beds, looking "like the vertebrae of some great sea serpent." Edwin Bryant described passing "immense piles of rocks, red and black, sometimes in columnar and sometimes in conical and pyramidal shapes, thrown up by volcanic convulsions.[4] These, with deep ravines and chasms, and widespread sterility and desolation, are the distinguishing features of the landscape."

The trail passes through the eroded cores of anticlines—huge arching folds of rock strata, like a stack of magazines bowed up in the middle—and through synclines, the reverse of anticlines, where the strata bow down in the middle. Lonesome pump jacks bob gracefully on the sagebrush hills. They are parked over anticlines, sucking up the crude oil trapped in the bowed-up layers. Oil underground percolates upward until it runs into rock that stops it. Rising droplets of oil collect against impermeable rock layers within the arching anticlines, forming caches of black gold. More than anything, oil has put Wyoming on the map. And more than any other rock formation, the Mowry Shale has put the oil in Wyoming. You can't miss the Mowry. It ranges dusk-gray to night-black from organic residue and smells faintly of decay. To oil geologists, it

4. Although volcanism was a favorite emigrant explanation for deformed (bent, broken, or tilted) rock, the distorted sedimentary layers so evident around the Foreland Ranges were caused by sideways squeezing of the Earth's crust during the Laramide Orogeny. More about this in chapter 7.

smells of money. The Mowry Shale formed as fetid mud on a Cretaceous seabed. Algae and plankton dying in the surface waters sifted down to the oxygen-deprived seafloor and accumulated, safe from oxygen-dependent microbes. Buried eventually several thousand feet underground, and cooked gently by the Earth's heat through the ages, the organic goop turned into crude oil. You cannot easily pump oil from the Mowry; it would be like trying to suck the oil through a dinner plate—the shale is too dense and impermeable. But eons of time and megatons of pressure have squeezed millions of barrels of Mowry crude into more porous and permeable rock neighbors—like the Frontier Formation, a layer of ancient beach sand that lies right above the Mowry. The permeable sand surrenders the oil quickly to a hungry pump jack. Mowry Shale crude, pumped from Wyoming and cracked in refineries, has poured into American gas tanks for decades.

Fifty miles and three days' travel west from the North Platte, the emigrants reached the banks of the Sweetwater River. Here they had their first taste of snowmelt from the Wind River Mountains, 100 miles away on the Continental Divide. The valley of the Sweetwater points like an arrow west to South Pass—the halfway point of the journey. "We began to ascend it with renewed spirits," Sarah Royce recalled, "knowing that when we reached its head we should soon pass the summit of the Rocky Mountains."

TO THE BACKBONE
OF THE CONTINENT

A few miles upstream from where they gained the Sweetwater River, the emigrants came to Independence Rock—a half-mile-long ridge of granite that rises like a whaleback from the sagebrush sea of the Sweetwater Valley.

It was a rite of westward passage to write one's name on Independence Rock. Thousands of signatures plastered its surface during the emigration years, and many remain today. Emigrants carved their names with hammers or chisels, or painted them with sticky mixtures of black powder and buffalo grease.

> After breakfast, myself, with some other young men, had the pleasure of waiting on five or six young ladies to pay a visit to Independence Rock. I had the satisfaction of putting the names of Miss Mary Zachary and Miss Jane Mills on the southeast point of the rock, near the road, on a high point. Facing the road, in all splendor of gunpowder, tar and buffalo greese, may be seen the name of J.W. Nesmith, from Maine, with an anchor. (J. W. Nesmith, July 30, 1843)

A party of 1830 trappers and traders lead by William Sublette probably gave Independence Rock its name when they celebrated Independence Day at its base. For the emigrants, the name signified westward progress. If you reached Independence Rock by the Fourth of July, you were on schedule to get over the western mountains before winter snows. Many emigrants climbed the 100 feet to the top of the rock, as do hundreds of visitors today. Today's tourists usually climb the north end, closest to

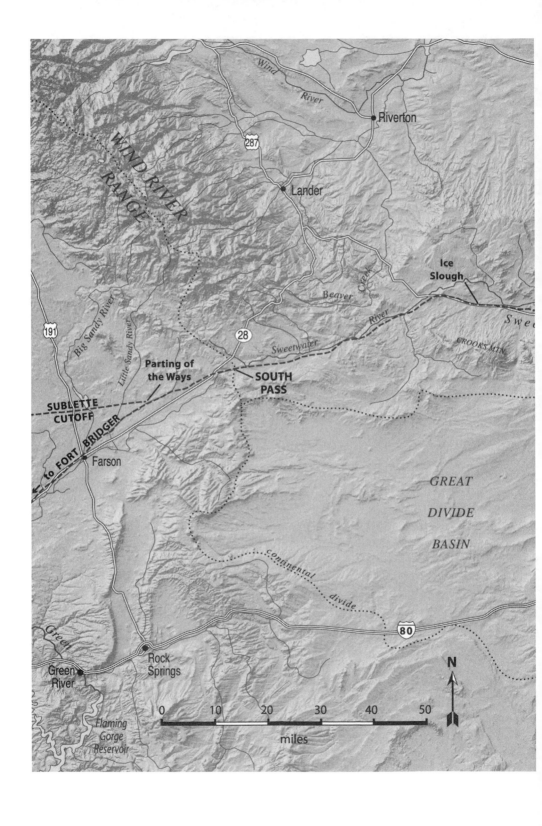

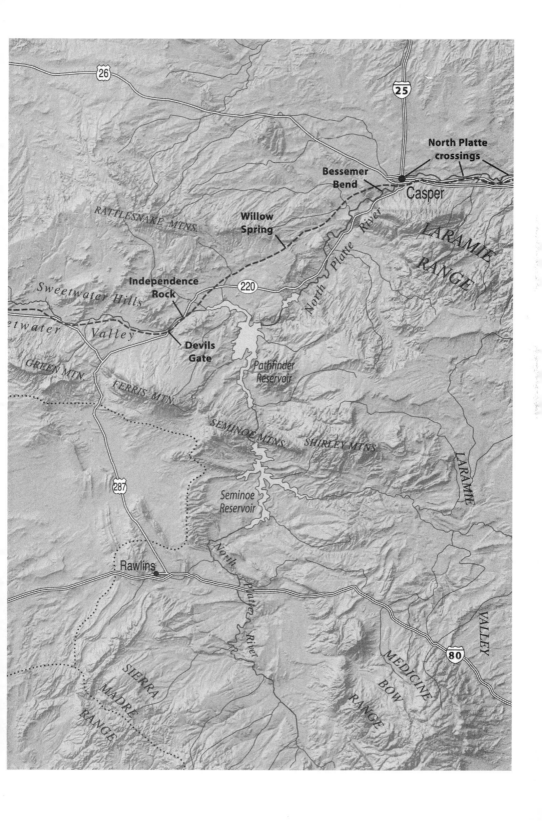

6.2 Emigrant names carved into granite on top of Independence Rock.

6.1 (*previous pages*) The Sweetwater Valley took the emigrants smoothly uphill 100 miles to the Continental Divide at South Pass. This fortuitous gap in the Rockies exists because of geologic happenstance. A Wind River–sized mountain range—the Sweetwater Range—once filled the east-west gap where the Sweetwater Valley is now. Several million years ago, it foundered to form the valley, thus opening the way west to South Pass. The Sweetwater Hills represent the exposed ridgeline of this buried range. These granite hills include two of the most famous landmarks on the Oregon-California Trail: Independence Rock and Devils Gate.

Wyoming Highway 220. The Sweetwater River flows past the south end, though, and this is probably where most emigrants climbed since this is where they would have first reached the rock. I shared the south end one June evening with no one but a red fox. We looked out on the green bottomlands of the Sweetwater, riotous with wildflowers, the stream twisting past in silvery loops. A squadron of ducks approached and landed in a hiss of spray. Beyond the skinny river, dusty sagebrush plains stretched off toward distant, cracked mountains. The sun landed on the horizon and sprayed everything with gold. Only a cigar could have improved that moment. Only a beer could have improved the cigar. Alas, my wagon was back at the highway.

INDEPENDENCE ROCK is made of broken, peeling granite. The rock is ancient even by geologic standards: more than 2.6 billion years old,—more than half the age of the Earth itself. Back then, the Earth's surface was nothing but rock, wind, and rain. No living thing existed on land, and life in the oceans had progressed only to the stage of single-celled slime. The rock is part of the Sweetwater Hills, which stretch west for 50 miles along the course of the Sweetwater River. The granite of Independence Rock and the Sweetwater Hills belongs to the so-called 2,600-Million-Year Age Group—a collection of ancient granites that cores large portions of the Laramie Range, Wind River Range, and other Wyoming mountains. The 2,600-Million-Year Age Group is part of the basement rock of the Wyoming Province—one of the great continental blocks assembled long ago to become North America. (Figure 3.5 shows the Wyoming Province in relation to the other basement blocks of North America.) Back when the south edge of North America ended in southeastern Wyoming—back when there was nothing but deep blue sea where Colorado and everything south of Colorado is today—the granite of Independence Rock was already 800 million years old.

Close up, the granite looks like a mix of cinnamon, salt, and pepper—a dazzling array of quartz, plagioclase, orthoclase, and mica minerals. Marble-sized, golf-ball-sized, even fist-sized phenocrysts of plagioclase feldspar poke up brusquely from the surface, standing high while the surrounding matrix crumbles away. Granite exposed to water and weather disintegrates from the outside in. The reason is mica—a mineral that forms booklike sheaves that easily peel apart. Mica is chemically weak; water converts it rapidly into clay. As the mica disintegrates, the tougher quartz, orthoclase, and plagioclase crystals fall out like bricks from a crumbling wall.

Granite begins as molten rock, 1,800 degrees hot or hotter. The transition from liquid magma to solid granite involves a slow process of atomic

6.3 Independence Rock. Thousands of emigrants scratched or painted their signatures on this celebrated granite landmark. The rock is about one-third of a mile long and over 100 feet high. The view looks east, and the Sweetwater River flows past the south (*right*) end. The arrow points approximately to the spot where J. Goldsborough Bruff sat when he drew the sketch shown in figure 6.4.

assembly. When hot, the atoms in magma—silicon, oxygen, sodium, aluminum, calcium, magnesium, iron, and others—ricochet wildly around in the soup of molten rock. But with cooling, they calm down. They begin to link arms, chemically speaking, to form crystals. Atoms of opposite electrical charges join in ionic bonds. Atoms that have electrons to share with others couple in covalent bonds. One by one, they assemble into precise geometric arrangements dictated by their sizes, with smaller atoms fitting into the spaces between larger ones. They line up in rows, columns, and three-dimensional networks. They join up by the hundreds, thousands, millions, and billions as the magma congeals, growing into sizable crystals.

Only deep underground does magma cool slowly enough to make the large, handsome crystals that make up granite. Turn a piece of granite in the light, and the crystal faces and cleavage planes flash at you, showing off their luster, their fine lines assembled with atomic precision. Magma that erupts to the surface as lava solidifies quickly, and the atoms have time only to form small crystals. Pick up a piece of basalt from a lava flow and run your fingers over the surface; the sandpaper texture shows how big the crystals were able to grow in the short time it took the lava to congeal—crystals not much bigger than the period that ends this sentence. But a stately pace of cooling deep underground gives the atoms time to assemble into crystals as big as fingernails or fingers or even fists.

Walk on granite, and you walk on rock that formed perhaps 10 or more *miles* underground. An astonishing fact, but no more so than what has to happen next. For you to touch granite, a stunning amount of geologic work must first be done. The granite must be uplifted from miles down, where it formed, and the intervening rock overhead removed. (This is

Henry Austin, M.D.
John Bates
Chs. G. Moxley
J. Goldsborough Bruff, Capt.
Washington City Company
July 26, 1849.

why you find granite in the uplifted cores of so many mountains, including the Rocky Mountain Foreland Ranges.) The granite decompresses as erosion scrapes away the overlying rock. The release of pressure produces distinct expansion cracks called joints. Joints in granite often come in sets of concentric, curving planes. The broken rock peels away from these cracks like the layers of an onion. Segments of these multi-ton exfoliating slabs lie on the sloping sides of Independence Rock, primed to slide off and join others collected around the base.

VIRTUALLY BACK to back with Independence Rock, six miles to the west, lies Devils Gate. Here, in an archetypal display of stream superposition, the Sweetwater River has cleaved a 300-foot-deep chasm through a granite ridge in the Sweetwater Hills (fig. 5.5). Devils Gate moved more emigrants to descriptive rhapsody than any other geologic landmark except perhaps Chimney Rock or Scotts Bluff. "It is grand, it is sublime! He must be brainless that can see this unmoved," John Edwin Banks declared. Lucy Cooke thought it "a grand sight! Surely worth the whole distance of travel."

> The "Sweetwater" rushes through an opening in the rocks, the walls on each side rising several hundred feet perpendicularly, and as though riven in two by some great convulsion of nature. . . . Oh, it is a most wonderful and sublime sight. . . . I bathed dear little Sarah's feet in the rushing waters, and only wish she had been able to realize the grand occasion.

The Sweetwater Hills, home to Independence Rock and Devils Gate, run east-west along the course of the Sweetwater River. The ancient granite forms a checkerboard of cracks—joints that formed, probably, during the same decompression process that cracked Independence Rock. Water gets into the joints and rots the rock. Tough trees like junipers gain a beachhead on bare granite by planting their roots into water-softened joints. The roots grow like wedges, assisting water in the work of cracking the face of the granite mountains into countless rectilinear blocks.

THE WAGON trains rolled on, past Independence Rock and Devils Gate, snaking west along the gently rising Sweetwater Valley. "The road can be

6.4 (*previous pages*) J. Goldsborough Bruff's sketch from the top of Independence Rock. The view looks southeast across the bottomlands of the Sweetwater River. The corralled wagons and approaching trains were a near-constant feature here in emigrant days. At the bottom of the sketch, Bruff wrote, "View from the top of Independence Rock.—looking to the S.E. [southeast] exhibiting the Sweetwater river & mountains:—the Washington City Comp. [Bruff's party] is corralled below. noon July 26, 1849." Reproduced by permission of the Huntington Library, San Marino, California.

6.5 These modern jeep tracks follow the original Oregon-California Trail up the wide Sweetwater Valley. The view looks west toward South Pass, beyond the horizon about 40 miles ahead. The river (not visible) lies several miles to the north (*right*). The broad, gentle rise of the Sweetwater Valley gave the emigrants an easy ascent to the Continental Divide at South Pass.

seen winding over the plains for many miles like a great serpent," W. S. MacBride marveled, "and train after train of white-topped wagons can be seen either way as far as vision can extend; and then beyond the scope of vision dust can be seen rising in the clouds." Not everyone along the Sweetwater was westbound, as Joseph Middleton saw in July 1849. "A Mormon from the Salt Lake settlement passed today with a wagon. He is going eastward to pick up old iron or anything valuable that he can find along the trail and take it back. He will not have to go far before he gets his wagon filled." The strain of the journey continued to take a grisly toll on livestock. Byron McKinstry found the road along the Sweetwater "well lined with carcasses and the smell is anything but agreeable."

The bottomlands of the Sweetwater River form a band of exuberant green winding through a parched land of gray and beige. The transition from the green bottomlands to the sagebrush plains and rock-strewn alluvial slopes of the open valley is a biological knife-edge. Within the water zone, life twitters and rustles as noisy as an Everglades swamp. Bushy 20-foot willows crowd the banks, hosting multitudes of redwing blackbirds. Ducks ply the channels and sloughs. Flowers bloom, insects buzz, and life goes about its busy business. Step away from that thin green line and you enter a realm of dust, saleratus, lizards, antelope, and sagebrush. The silence would be almost complete on these arid plains, if not for the wind.

Wind—tyrant king of Wyoming. Wind-driven sand blasts the paint off buildings and gnaws out the softer layers of wood between the harder growth rings. For most houses on the exposed plains, a sheltering band of trees planted on the west side is as essential as a roof. Without the trees, you could lose the roof. Or your mind. The wind sculpts the trees into misshapen weathervanes, streaming east. Wyoming highway rest areas have wind shelters made of two high brick walls that join at a V, like a ship's prow. The V points west as predictably as a compass points north. The Wyoming tourist board is in denial. *Come tourists, enjoy a picnic in our lovely highway rest areas. Just bring along some bolts to fasten your sandwiches to the table.* Fifty-pound sandbags weight the bottoms of the rest-area trash cans. Without them, the big steel cans bounce away like Styrofoam cups. Forlorn cows endure a lifetime of wind, joylessly converting the sparse grass of the plains into meat until slaughter brings relief. Legends tell of people driven to murderous insanity by the wind.

"We traveled late and encamped in a driving storm of wind, which took up the sand and gravel and carried it along like shot," William Swain recorded one July night along the Sweetwater. "The wind, in addition to its furious violence, is so very hot and dry as to render respiration . . . quite difficult," Howard Stansbury complained. "The throat and fauces become dry, and lips clamy and parched, and the eyes much inflamed from the drifting dust."

Wyoming is nearly as generous with snow as with wind. The state ranks dead last in U.S. population, but it may rank number one in total mileage of highway snow fences. These tall, slanted walls of slatted boards stretch for miles along the roads, facing west. In the epic winter battles between nature and the highway department, snow fences are the first line of defense. Snow never melts in Wyoming, the local saying goes—it just blows back and forth until it wears out. Snow can come any day of the year, as the emigrants learned. "At dark, while I was cooking supper, a heavy storm of wind and snow came up," Margaret Frink wrote in June 1850, while camped near Willow Spring. "There was no shelter, and we ate our supper while it was snowing and blowing." The next day she and her husband and companions "snowballed each other till ten o'clock, when the sun got too warm for the snow to remain." The novelty of snow in the middle of summer triggered many an emigrant snowball fight. "We found a bank of snow some 2 to 20 feet deep," James Pritchard reported one June day along the Sweetwater, and added, "The Boys took quite an exciting game of snow bawling in which a dozen or more took part."

Notwithstanding wind and snow, the gentle ascent up the broad valley of the Sweetwater was rather easy, even monotonous. About 40 miles west of Devils Gate, though, the monotony was interrupted by a great curiosity. The trail crosses a band of marshy land several hundred yards wide made by a tributary of the Sweetwater. Here, about one foot below the surface, the insulating cover of peat and plants preserved—even in the height of summer—a layer of solid ice. James Pritchard found the ice "clear and pure, and as good as any I ever cut from the streams in Kentucky. I cut and filled my water bucket & took it to camp with me." Few emigrants passed Ice Slough[1] (also called Ice Spring) without stopping to dig for ice, "so desirable a luxury in a march so dry and thirsty."

THE RISING plain west of the final crossing of the Sweetwater River is so gentle that many emigrants scarcely noticed where the trail leveled out, and then, ever so subtly, the gullies beyond drained west. They had reached the Continental Divide, South Pass, nearly 1,000 miles from the Missouri River, halfway to California. "We could hardly realize that we were crossing the great backbone of the North American continent," Margaret Frink marveled.

> The ascent was so smooth and gentle, and the level ground at the summit so much like a prairie region, that it was not easy to tell when we had reached the exact line of the divide. But it is here that after every shower the little rivulets separate, some to flow to the Atlantic, the others into the Pacific.

South Pass, 7,550 feet above sea level, was the great tipping point on the westward journey. Here, the division of waters, Atlantic versus Pacific, signified the cleaving of one's life. The past lay east, the future west—and for many emigrants, there would be no returning. It was a moment mixed with excitement, anticipation, and regret, as Sarah Royce remembered:

> I had looked forward for weeks to the step that should take me past that point [the divide]. In the morning of that day I had taken my last look at

---

1. Ice Slough lies along Wyoming Highway 287 a few miles west of Jeffrey City. There is little ice in summer there now, possibly because of global warming.

the waters that flowed eastward, to mingle with the streams and wash the shores where childhood and early youth had been spent; where all I loved, save, O, so small a number lived; and now I stood on the almost imperceptible elevation that, when passed, would separate me from all these, perhaps forever.

Despite its weighty symbolism, South Pass is a visual anticlimax—a monotonous high plain coated with haggard sagebrush and sparse grass. The pass is saddle-shaped, rising up to low hills on the north and south, and bowing gently down to the east and west. The land is treeless. No plant grows much over a foot tall; all hunker before the wind. South Pass is not a welcoming place. "There was a gloomy vastness in the distant prospect and a sense of loneliness that was really dispiriting," A. J. McCall wrote on his arrival on July 5, 1849.

Nonetheless, gaining South Pass was the emigrants' best excuse yet for a party. Peter Decker "got out the Star Spangled Banner and planted it on the South Pass. A breeze waved, our folks met around it and passed a cheerful evening fiddling, singing and dancing on a sheet of zinc." A midday hailstorm did not stop Margaret Frink's party from partying. "Music from a violin with tin-pan accompaniment, contributed to the general merriment of a grand frolic." Dr. Charles Parke took milk from his cows and, using snow, buckets, and salt, made ice cream—"the most delicious ice cream tasted in this place," he claimed, and fairly too.

Mounting the low hills near South Pass gives you an unbroken view across some of the emptiest land anywhere in the United States. Other than one distant paved road (Wyoming Highway 28), the view today hardly differs from the one Edwin Bryant took in on July 12, 1846.

> Just before sunset, accompanied by Jacob, I ascended one of the highest elevations near our camp; we took a farewell look of the scenery towards the Atlantic. The sun went down in splendor beyond the horizon of the plain, which stretches its immeasurable and sterile surface to the west as far as the eye can reach. The Wind River Mountains lift their tower-shaped and hoary pinnacles to the north. To the east we can see only the tops of some of the highest mountain elevations. The scene is one of sublime and solemn solitude and desolation. The resolution almost faints when contemplating the extent of the journey we have already accomplished, and estimate the ground which is yet to be traveled over before we reach our final destination on the shore of the Pacific.

Laid out in every direction from South Pass sprawls a landscape unearthed and carved up during the Exhumation. To the northeast, the land plunges over the Beaver Divide into the dissected headwaters of the Wind

River drainage. North-flowing Beaver Creek is cutting headward, south, into the divide. Its headwater ravines are scratching like claws toward the Sweetwater River, only five miles south. In the millennia ahead, Beaver Creek's ravines will cut south all the way to the channel of the Sweetwater and capture it. Stream piracy.

To the north and northwest, the Wind River Range raises its "cold, spiral and barren summits" to the clouds. The Wind River may be the wildest and most remote mountain range in the lower 48 states. One paved road wraps around its southeast end. The next one crosses the mountain 140 miles to the northwest. There is nothing in between but rock, snow, and wild alpine forest. The range's 100-mile-long ridgeline is the Continental Divide, separating Missouri River drainage from Colorado River drainage. The north end of the range drains to the Columbia River, making the Wind River Range the only mountain in North America that drains to all three master streams of the American West: the Missouri, the Colorado, and the Columbia. Nearly a dozen Wind River peaks reach above 13,000 feet, including Wyoming's highest point: 13,804-foot Gannet Peak. On the southwestern flank of the range, at 10,000 feet, lies a dissected subsummit surface, more than five miles wide—a palimpsest of the pre-Exhumation Rocky Mountains (shown in fig. 5.7).

Turning west, the view takes in the immense Green River Basin—a 60-mile-wide sagebrush plain that the emigrants crossed west of South Pass. Its apparent flatness belies the countless arroyos and deep canyons that the Green River and its tributaries continue to carve as they carry on with the Exhumation. The Green River exits the basin to the south, through deep canyons superposed through the Uinta Mountains in Utah, just across Wyoming's south border. The Yampa River joins the Green in the heart of the Uintas; two rivers that flow into a mountain and join at its center—an impossibility made possible by stream superposition and the Exhumation.

South and southeast of South Pass, the land is buckled into gentle arches and basins. Fins and ridges of hard, competent strata form miles-long hogbacks where they arch over the Rock Springs Uplift and the Rawlins Uplift. The Continental Divide splits in two here to go around the Great Divide Basin—a vast, antelope-dotted bowl of rolling grassland 70 miles across. The Great Divide Basin is a true topographic oddity, a depression perched on the Continental Divide. Rain in this part of Wyoming, depending where it lands, might flow to the Atlantic, to the Pacific, or pool up in the Great Divide Basin.

Turning east to complete the circle, the view sweeps down the axis of the Sweetwater Valley. Nowhere else in the Rockies does a river valley rise so smoothly and steadily to such a low crossing of the Continental

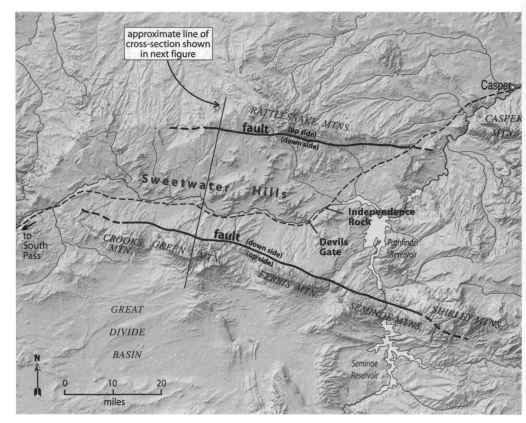

6.6 Shaded relief map of the Sweetwater Valley. A mountain range that I call the Sweet-water Range once lay where the Sweetwater Valley is today. The Sweetwater Hills are its exposed tips; the rest of the mountain lies buried below the valley floor. The Sweet-water Valley formed when the Sweetwater Range foundered like a wedge between large faults on its north and south sides. The original foothills of the Sweetwater Range are represented by small mountains that lie north and south of these faults (the Rattlesnake Mountains to the north, and Crooks, Green, Ferris, and Seminoe mountains to the south). These remnant foothills were left standing high as the Sweetwater Range slid down to form the valley. Figure 6.7 illustrates this sequence of events along the cross-section line above.

Divide. It was not always so. The Sweetwater Valley gateway to South Pass exists by a stroke of geologic good fortune.

Today the Sweetwater Valley stretches 100 miles between the Wind River Range and the Laramie Range, forming an obliging gap through the Foreland Ranges. More than 400,000 souls took advantage of this gap to pour west in the mid-nineteenth century alone. But when the Wind River and Laramie ranges originally rose, a third range—we can call it the Sweetwater Range—came up with them in between. Later

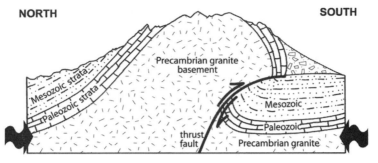

NORTH   **SOUTH**

Precambrian granite basement

Mesozoic strata

Paleozoic strata

Mesozoic

Paleozoic

thrust fault

Precambrian granite

~60 million years ago: Sweetwater Range rising from sideways compression during the Laramide Orogeny

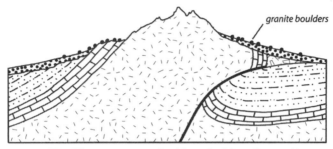

granite boulders

~35 million years ago: Boulders shed from the granite core of the range pile up in the foothills.

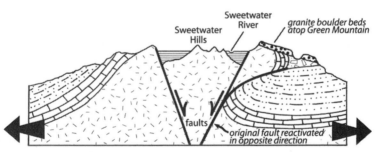

Sweetwater River

Sweetwater Hills

granite boulder beds atop Green Mountain

faults

original fault reactivated in opposite direction

~10 million years ago: Sideways stretching of the crust drops the range to make the Sweetwater Valley.

6.7 The rise and fall of the Sweetwater Range. The range originally rose along with the rest of the Rocky Mountain Foreland Ranges in response to sideways compression during the Laramide Orogeny. Erosion shed granite boulders from the uplifted core of the range into its foothills. Then, about 10 million years ago, sideways stretching of the Wyoming crust (possibly related to crustal stretching in the Basin and Range Province or in the Rio Grande Rift of New Mexico) caused the range to slide back down faults that developed on its north and south flanks. (The south fault appears to be, in part, the original fault along which the range squeezed upward. As the crust stretched, the mountain slid back down the same fault.) The granite boulders today capping Green Mountain and Crooks Mountain bear evidence to the now-vanished heights of the Sweetwater Range.

the Sweetwater Range foundered like a great wedge between two faults, forming the Sweetwater Valley. The exposed tips of the foundered range form today's Sweetwater Hills, poking up from the sand and gravel of the Sweetwater Valley like the roofline of a house buried in an avalanche. The rest of the Sweetwater Range is out of sight, below the winding course of the Sweetwater River.

Imagine the geography of Wyoming if the Sweetwater Range were still standing. A continuous mountain barricade would zigzag northwest across the state: Laramie Range–Sweetwater Range–Wind River Range–Absaroka Range. There would have been no Sweetwater Valley, no easy gateway to South Pass. The Oregon-California Trail would not exist as we know it.

The foothills of the Sweetwater Range still stand high today. Because these foothills lay outboard of the faults that dropped the center of the mountain, they stayed put while the rest of the mountain fell away. Today they make up the Rattlesnake Mountains to the north of the Sweetwater Valley and four small mountains called Crooks, Green, Ferris, and Seminoe to the south (fig. 6.6). Crooks Mountain and Green Mountain present steep faces north, toward the Sweetwater Valley, and planar summits that slope gently south, like buildings with one-sided roofs. Capping those roofs are thick beds of granite boulders, about 35 million years old. Traced uphill to the north, the boulder beds end, hanging out into empty space at the mountaintops. It seems impossible—granite boulders do not fall from the sky. But when you project the sloping boulder beds north into the air over the Sweetwater Valley, the south flank of the vanished Sweetwater Range materializes in the sky. It's a beautiful moment—a eureka moment—when you stand on the boulder beds on top of Green Mountain and understand. Time shifts instantly to the lofty days of the Sweetwater Range. You watch as the granite boulders come whistling and bouncing down from the now-vanished heights, rolling to a stop in the foothills around you. Now those foothills, orphaned by the collapse of their mountain mother, perch on the south edge of the Sweetwater Valley and look *down* on the Sweetwater Hills—the source of the boulders that once rolled down from on high (fig. 6.7).

THIS GLANCE at the rise and fall of the Sweetwater Range cracks the door onto a multimillion-year history of mountain upheaval in the American West. When continents move across the face of the Earth, mountains rise along their leading edges. North America is a continent on the move. For most of the past 200 million years, the continent has been sliding west and overriding ancient seafloor. Mountains have blistered upward along its western edge in response. We turn now to this westward journey—the journey of a continent and the mountains that it spawned.

CORDILLERAN UPHEAVAL

Open a good atlas to the pages that show North America and a singular division leaps out at you. The eastern two-thirds of the continent lie flat, like a freshly made bed with a few wrinkles. The western third is like a bed after a restless sleep, the covers pushed aside into a rumpled heap. A wrinkled landscape of mountains stretches from Alaska to Panama along the continent's west flank. This is the North American Cordillera—one of the world's great mountain belts.

California-bound emigrants gazed up at Cordilleran peaks for three months and 1,300 miles of their journey, all the way from Wyoming's Laramie Range to the Sierra Nevada. "You may think you have seen mountains and gone over them," William Wilson wrote in a letter home to his Missouri kin in 1849, "but you never saw anything but a small hill compared to what I have crossed over, and it is said the worst is yet to come." South Pass—the dividing line of the continent and the halfway point of the emigrant journey—is an apt place to leave the emigrant trail, for now, and explore the story of these mountains.

Mountains exist because the Earth's surface is divided into moving plates of rock, hundreds to thousands of miles across, 50 to 100 miles thick. The plates cover the Earth like a fractured eggshell, and they slide across the face of the Earth in various directions a few inches each year. That's a plodding pace in human terms, to be sure, but it's enough to add up to thousands of miles of travel over the scope of geologic time. The plates crunch into each other at some of their shared edges, pull apart at others, or grind side by side past one another. They move because the Earth, as it has for eons, is slowly releasing its immense reservoir of internal heat. The outward flow of heat from the bowels of the planet triggers convective motion of hot rock in the mantle below the plates. Areas of hotter rock rise up; areas of less hot rock sink down. The movement of the plates is the outward expression of this internal heat-emitting engine.

Continents ride like passengers atop the plates, going where the plates take them. The edges of the plates—the boundaries where one plate ends and another begins—lie mostly in the ocean basins. For instance, the South American Plate begins at the Mid-Atlantic Ridge, halfway across the South Atlantic. It extends west 3,500 miles, clear across South America to the Peru-Chile Trench off the continent's west coast. From there,

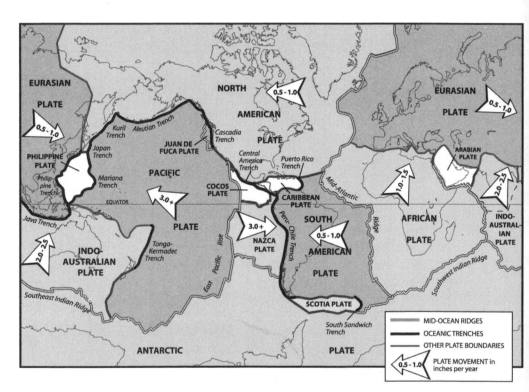

7.1 The Earth's major tectonic plates. The edges of the plates lie mostly in the ocean basins, forming mid-ocean ridges and oceanic trenches. The continents ride on the plates, going where the plates take them. The Atlantic Ocean and the Rea Sea are widening by seafloor spreading at mid-ocean ridges, carrying continents apart. The Pacific Ocean, the Mediterranean, and the Persian Gulf are closing by subduction, and mountains are rising on the adjacent continents in response. The arrows show the rates and directions of plate movement, as determined by satellite measurements. The Mercator map projection stretches out the polar regions, making the Antarctic Plate and North American Plate appear larger than they actually are.

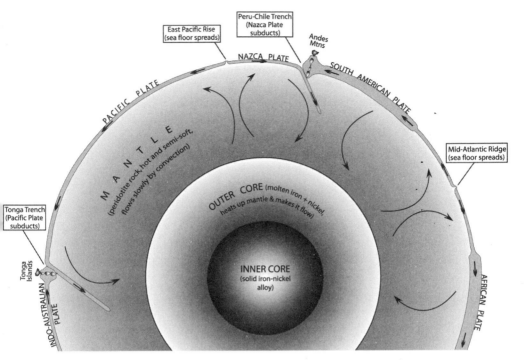

7.2 A slice through the Earth showing its major layers. The slice bisects the Indo-Australian, Pacific, Nazca, South American, and African plates. For clarity, the diagram exaggerates the thickness and topographic relief of the plates. The oceans are too thin to show at this scale. The arrows depicting mantle convection are purely schematic; the actual pattern is not well known.

the Nazca Plate stretches 2,000 miles to the East Pacific Rise. There the Pacific Plate begins, extending nearly one-third of the way around the planet to the ocean trenches off Japan, Kamchatka, and the Aleutian Islands. Many plates—such as the South American, North American, and African—contain both continental and oceanic regions. Others, like the Nazca Plate and the Pacific Plate, carry no continents; they contain oceanic islands or no landmasses at all.

Where plates separate, continents drift apart and new ocean floor forms in between. The ocean floor grows from mid-ocean ridges—huge undersea mountain chains with rift zones running down their centers. Molten lava erupts in the central rifts and solidifies into basalt rock, forming new seafloor. Older seafloor makes room for new by spreading away from mid-ocean ridges like two oppositely moving conveyor belts—a process called seafloor spreading. About 250 million years ago, North American and South America butted up against Africa and Eurasia as part of a single great continent called Pangaea. A glance at the Atlantic shows the jigsaw puzzle fit of the South American and African coastlines. The Atlantic Ocean was born as North America and later South America tore slowly away from their moorings against Eurasia and Africa to form a mid-ocean ridge—the Mid-Atlantic Ridge—in between. Every year the Mid-Atlantic Ridge cranks out two to three inches of new seafloor as it spreads along its length, pushing New York incrementally away from Lisbon, and Rio de Janeiro away from Cape Town. Snug the continents on either side of the Atlantic back together, and rock bodies and mineral belts match up like lines of print across a piece of torn newspaper.

Where plates converge, ocean basins close, continents collide, and mountains rear up. India, riding on the Indo-Australian Plate, has collided with the southern edge of Eurasia. Its arrival closed the ancient Tethys Seaway and pushed up the Himalayas. Africa is creeping toward Eurasia, slowly collapsing the Mediterranean Sea and pushing up the Pyrenees, the Alps, the Zagros, the Caucasus, and other mountains in the process.

Mountains rise wherever plates bring continents into collision, but that's not the only way the Earth makes mountains. No India-sized or Africa-sized landmass has landed on the west coasts of North or South America. Yet there lay the two greatest mountain belts in the Western Hemisphere: the Cordillera and the Andes. These mountains are the offspring of subduction—the process whereby an oceanic plate dives beneath the edge of another plate. Deep troughs on the seafloor called oceanic trenches mark out these places, also called subduction zones. Oceanic trenches, or subduction zones, rim most of the Pacific Ocean. They make a great ring of volcanic violence—the Pacific Ring of Fire—that nearly encircles the Pacific basin. The Aleutian Trench, Kuril Trench, Japan Trench, Mariana Trench, and Kermadec-Tonga Trench all mark out places where the huge Pacific Plate, lumbering northwest, is subducting around the northern and western edges of its basin. The Peru-Chile Trench, Philippine Trench, Ryukyu Trench, and Java Trench likewise demark seafloor subduction zones.

All oceanic trenches, without exception, partner with volcanic arcs—lines of active volcanoes that run parallel to the trench. You can't have an oceanic trench without a parallel volcanic arc, because subduction makes them both. The downward bending of the seafloor as it plunges into the mantle forms the trench; the demise of that seafloor, coughing up magma as it slides to its hot death in the mantle, makes the volcanic arc. (Magma is more buoyant than the rock around it, so where it forms, it rises. It forces its way upward through fractures in the overlying rock or melts its way through like a blowtorch through wax.) The lurching movement

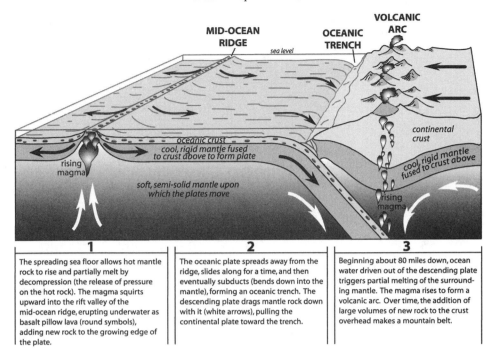

MID-OCEAN RIDGE

OCEANIC TRENCH

VOLCANIC ARC

sea level

oceanic crust
cool, rigid mantle fused to crust above to form plate

rising magma

soft, semi-solid mantle upon which the plates move

continental crust

cool, rigid mantle fused to crust above

rising magma

1 The spreading sea floor allows hot mantle rock to rise and partially melt by decompression (the release of pressure on the hot rock). The magma squirts upward into the rift valley of the mid-ocean ridge, erupting underwater as basalt pillow lava (round symbols), adding new rock to the growing edge of the plate.

2 The oceanic plate spreads away from the ridge, slides along for a time, and then eventually subducts (bends down into the mantle), forming an oceanic trench. The descending plate drags mantle rock down with it (white arrows), pulling the continental plate toward the trench.

3 Beginning about 80 miles down, ocean water driven out of the descending plate triggers partial melting of the surrounding mantle. The magma rises to form a volcanic arc. Over time, the addition of large volumes of new rock to the crust overhead makes a mountain belt.

7.3 The plate tectonic system is driven by the opposing processes of seafloor spreading and subduction. Plates grow by seafloor spreading at mid-ocean ridges. Plates converge, with one subducting (plunging down beneath the other) at oceanic trenches. The subducting plate triggers melting of the mantle, producing an arc of volcanic mountains running parallel to the trench.

of the seafloor as it goes down the trench also makes earthquakes. The Pacific basin, surrounded by subduction zones, has the greatest concentration of active volcanoes and earthquakes on the planet.

Subduction spawns orogenesis—mountain building—and there's no better place to see that in action than in the Andes. Subduction is pushing up the Andes even as you read this.[1] Temblors rattle the region daily as the rocks bend, break, and ooze up into mountains. Magma roils deep down and periodically erupts from the dozens of volcanoes that run like a row of Chinese hats along the 4,400-mile-long mountain belt. Both the North American Cordillera and the Andes are the offspring of subduction, but in the Cordillera subduction is now largely over, while in the Andes subduction continues. As we transit the Cordillera in the next part

1. When Charles Darwin visited the coast of Chile on his *Beagle* voyage (1831–36), he was astonished to find beds of seashells jacked up as much as 1,300 feet above sea level along the Andean coastline. He surmised (correctly) that the Andes are actively rising.

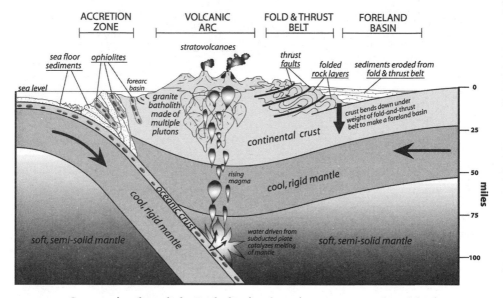

7.4 Cross section through the Earth showing the major components of a subduction-generated mountain belt, such as today's Andes and—for much of its history—the North American Cordillera. Compare this figure to the aerial view of the Andes shown in figure 7.5.

of this chapter, the model of the Andes will serve us well. The Andes present us with living examples of the same mountain-building processes that made the Cordillera millions of years ago—albeit with some key differences and complications.

The Andes begin, in effect, deep under water at the Peru-Chile Trench. Here the Nazca Plate, sliding east from a mid-ocean ridge (the East Pacific Rise), bends down beneath the west edge of South America and heads into the mantle. At the trench, the advancing west edge of South America acts like a plow blade, scraping deep-sea sediments off the subducting plate. Any large bumps that stick up from the seabed—abyssal hills,

7.5 (*on facing page*) Aerial image of the Andes between about 15 and 35 degrees south latitude. Subduction of the oceanic Nazca Plate at the Peru-Chile Trench makes the mountain belt. Note the parallel-trending volcanic arc, fold-and-thrust belt, and foreland basin, as portrayed in cross section in figure 7.4. Note also that the pattern is disrupted between 28 degrees and 33 degrees south latitude. Here the volcanoes are no longer active, while numerous basement-cored uplifts (the Pampean Ranges) have popped up within the foreland basin east of the volcanic arc. The reason appears to be flat subduction of the Nazca Plate in this area. A similar process may explain the uplift of the Rocky Mountain Foreland Ranges during the Laramide Orogeny. Background image courtesy of Dr. William Bowen.

0   100   200
miles

15 S

20 S

NAZCA

PLATE

PERU - CHILE TRENCH

25 S

Antofagasta

Copiapo

30 S

no active
volcanoes
between
~ 28 & 33
S latitude

VOLCANIC

ARC

FOLD

&

THRUST

BELT

FORELAND

BASIN

S. de Velasco

S. de Valle Fertil

S. de Cordoba

Sierra de Cordoba

Cordoba

S. de San Luis

Nazca Plate subducts flat in this zone

Valparaiso

35 S

Pampean Ranges
(basement-cored uplifts between
~ 28 and 33 degrees S latitude)

seamounts, or islands—may also be scraped off at the trench, joining the mishmash of sheared-off rock stuck to the west edge of South America. This is the tectonic accretion zone—the first of four major components of a subduction-generated mountain belt (see fig. 7.4).[2]

The Nazca Plate, robbed of some of its sedimentary blanket, angles down into the mantle. About 80 to 100 vertical miles down, the mantle's heat forces trapped seawater out of the subducting plate. The water leaks into the surrounding mantle, triggering melting of the hot rock and sending up masses of newly minted magma.[3] If you could see under South America, the magma would look like a slow-motion reverse waterfall of molten blobs, rising upward. The magma works its way upward into the core of the Andes. Much of it stalls and solidifies underground, far below the volcanoes at the surface. There it forms plutons—bloblike bodies of igneous rock up to several miles across. Multiple plutons fuse together to make batholiths—immense rock masses, tens of miles wide and hundreds long, formed of hundreds of amalgamated plutons. A batholith is like an old bag of marshmallows in which the individual marshmallows (plutons) have fused into a single hardened mass. Magma that doesn't stop underground bursts to the surface, making volcanoes. This is the volcanic arc—the second major component of a subduction-generated mountain belt. The mountains rise ever higher as newborn magma swells the crust with plutons and paves the surface with layer after layer of lava and ash.

As the volcanic arc swells and grows, the weight of the rock begins to exceed its own strength. The rock begins to slide down and flow away to the sides. (Visualize gobs of pudding dropped one after another onto

2. Not all subduction trenches experience tectonic accretion. If there isn't much sediment on the seafloor or in the trench to begin with, then little will scrape off and accrete. Much of South America's west coast is arid and steep, with small, short rivers that don't deliver much sediment to the Peru-Chile Trench. In many areas of the trench, the subducting Nazca Plate actually erodes the edge of the South American Plate as it grinds past on its way into the mantle—a process called tectonic erosion. Nonetheless, I emphasize the trench as a zone of tectonic accretion here because the accretion of rock and sediment at subduction trenches plays a big part in the evolution of many subduction-generated mountain belts—particularly the North American Cordillera, as we'll see.

3. Subducted water triggers melting of the surrounding mantle because—counterintuitive though it may seem—water lowers the melting point of rock. Experiments have shown that if you heat up mantle rock close to its normal melting point and add water, some of the rock melts. The effect is like scattering salt on an icy road; the salt interferes with the chemical bonds that form ice crystals, causing some of the ice to melt. Likewise, water molecules interfere with the chemical bonding between atoms in hot rock, so the presence of water causes the rock to melt at a lower temperature.

a plate. Each gob represents a pluton or a lava flow — new rock added to the volcanic arc. As the blobs stack up, the pile begins to flow outward.) Rock throughout the Andes responds by escaping to the east, away from the pressure. The rock bends into folds like a rumpled towel. It breaks into huge slabs that slide east along ramplike thrust faults. The bent and broken rocks pile up into a fold-and-thrust belt — a mountainous stack of distorted rock that runs parallel to, and inland of, the volcanic arc. (Notice the rumpled carpet look of the Andean fold-and-thrust belt in fig. 7.5.)

The fold-and-thrust belt creeps and oozes east onto the continental interior. Its colossal weight bows down the crust like a fat man on the end of a diving board, forming a foreland basin — a long, linear depression immediately east of the fold-and-thrust belt. This is the foothill country of the eastern Andes. Piled here is a monumental stack of sand and gravel — the eroded debris of the volcanic arc and the fold-and-thrust belt. The foreland basin would be a topographic depression were it not filled to overflowing with sand and gravel. The eroded debris piles up into a great wedge, thickest and highest next to the fold-and-thrust belt and sloping down to the east, away from the mountains.

That's the Andes in a nutshell, a mountain system born of subduction and built of four parallel components: tectonic accretion zone, volcanic arc, fold-and-thrust belt, and foreland basin (fig. 7.4). With the Andes as a living model, we can now transit the North American Cordillera in time and space and piece together its history — one that, like the Andes, harks back to subduction along the western edge of the continent.

THE NORTH AMERICAN CORDILLERA is a young mountain belt by geologic standards. Nearly all of the folding, faulting, and mountainous topography that you see from Wyoming to California is less than 200 million years old — 4 percent of the age of the Earth. We can split this interval into two major phases. The first phase, from about 200 million to 45 million years ago (Jurassic to Eocene time), was primarily one of sideways *compression* from the west, arising from subduction. The second phase, which began about 45 million years ago, was mostly one of sideways *extension,* in which the crust stretched in a generally east-west direction. In essence, the continent was first crushed from the west and then stretched to the west. Both processes made mountains, but the compression phase affected a much larger region. The compression phase squeezed up mountains all the way from Alaska to Mexico, while the extension phase was mostly limited to the Basin and Range, Rio Grande Rift, and the Sierra Nevada provinces — topics for later chapters.

The compression phase was the golden age of Cordilleran mountain building. Its effects are writ large across the American West. Vast amounts of new land, rafted in by subduction along the west coast, glommed onto the continent in the tectonic accretion zone, assembling most of California, Oregon, Washington, and British Columbia. Meanwhile, to the east, volcanoes ran riot as thousands of cubic miles of subduction-generated magma welled up to form a long-lived volcanic arc. Fold-and-thrust belt mountains shouldered up east of this arc. Advancing east across Nevada and Utah, the fold-and-thrust belt shed eroded debris eastward into a vast foreland basin—a world of rivers, forests, and swamps where dinosaurs roamed. East of there, for a time, a 1,000-mile-wide ocean—the Cretaceous Interior Seaway—ran right up the middle of the continent, above today's Great Plains, reaching from the Gulf of Mexico to the Arctic Ocean. The Rocky Mountain Foreland Ranges reared up from the muddy floor of this interior sea. They expelled the sea and reached for the clouds even as the clouds brought them down. Much later they rose again, rejuvenated during the Exhumation.

All of this mountain-building action burst forth because, for more than 200 million years, the North American Plate has been sliding west. The eastern edge of the plate—the mid-Atlantic Ridge—has grown by seafloor spreading as the Atlantic has widened. Heading west, the North American continent has overridden several thousand miles of ancient seafloor along its western edge. The story of Cordilleran upheaval therefore begins with North America's westward migration.

North America Heads West

Two hundred fifty million years ago, North America was locked in a mutual embrace with South America, Africa, and Eurasia. They were joined by India, Antarctica, and Australia, welded into a gargantuan continental mass—Pangaea. The west side of North America faced the ancient Pacific Ocean.

Pangaea's size was probably its undoing. Continental crust is thick and doesn't conduct heat efficiently. Huge Pangaea, more than 8,000 miles across east to west and stretching nearly from pole to pole, blocked much of the heat attempting to escape from the Earth's interior. The insulating effect was probably greatest right in the center of the supercontinent, under the thick welt of mountains raised by North America's collision with northwest Africa during the assembly of Pangaea. (That collision raised Himalayan-sized mountains whose deeply eroded roots form today's Appalachians.) The warming mantle under Pangaea began to expand and rise, spreading out inch by inch below the supercontinent like a colossal rock mushroom cloud. By late in Triassic time, about 220 million years ago, Pangaea was splitting at the seams.

Heading west out of Pangaea, North America began to override the ocean floor next door—a slab of seabed called the Farallon Plate. The plate bent down under the advancing western edge of the continent (like the Nazca Plate does under South America today), to form a subduction zone.

Subduction of the Farallon Plate built the North American Cordillera. Since the breakup of Pangaea, an expanse of the Farallon Plate roughly equal to the current width of the Pacific Ocean has disappeared beneath North America.[4] The subduction injected the continent with immense volumes of magma and threw up towering volcanic mountains. It also compressed the continent from the west, pushing up fold-and-thrust belt mountains and later squeezing the Foreland Ranges out of the basement.

As if that weren't enough, the Farallon Plate also added vast amounts of new land to North America's western edge. Recall how the edge of a continent at a subduction zone acts like a snowplow blade, scraping off seabed sediments, seamounts, islands, and even small continents carried into the trench. Occasionally, the continental blade digs a bit deeper into the subducting seabed, dislodging whole slabs of oceanic crust like a snowplow tearing out a chunk of roadbed. As newer material wedges in behind these dispossessed seafloor slabs, they are pushed up onto the continent to make new land. Throughout California's Coast Ranges and western Sierra Nevada foothills, you can drive for miles across slices of ancient abyssal seabed. These singular rock bodies, evicted from their ocean home during subduction of the Farallon Plate and marooned onto North America, are called ophiolites. They consist largely of contorted pillow basalt—the lava rock formations that make up the ocean floor.

4. We can make a rough estimate of how much of the Farallon Plate has been subducted under North America as follows. Throughout the world, the amount of new ocean floor manufactured by seafloor spreading at mid-ocean ridges must be balanced by the destruction of ocean floor at subduction zones. Since the time that North America broke free of Pangaea, it has moved west by about 1,700 miles—the distance from the Mid-Atlantic Ridge to the edge of the eastern U.S. continental shelf. In other words, if the Farallon Plate had stood still, North America would have overridden 1,700 miles of it. But the Farallon Plate did not stay still. Seafloor spreading along the ancient East Pacific Rise pushed the Farallon Plate *east* under North America at least two times faster than North America moved *west*. A conservative estimate of the total amount of Farallon Plate ocean floor subducted underneath North America is therefore 1,700 miles (North America's westward movement) + (2 × 1,700 miles) (the Farallon Plate's eastward movement) = 5,100 miles, or roughly the distance from San Francisco to Tokyo.

An ophiolite's journey begins on the deep ocean floor. As seawater percolates into hot, newly formed oceanic crust at mid-ocean ridges, the heated water alters the minerals in the rock to give it a smooth, fibrous texture and a mottled, sea-green luster. The result looks a bit like green-black snakeskin, giving the rock its name: serpentinite. Borrowing on the theme, geologists gave the name ophiolite (from *ophis*, Greek for snake) to slices of serpentinized seafloor extricated from the seabed and pushed up onto continents during subduction. The Farallon Plate's long history of subduction has beached many ophiolites on North America's west coast, particularly in California. Serpentinite is California's state rock.

Like groceries piling up at the end of a checkout-line conveyor belt, all manner of geologic real estate has collected against the western edge of North America, courtesy of the subducting Farallon Plate. Each piece, as it docked, added new land to the continent. Collectively, these chunks of imported rock are called accreted terranes.[5] Nearly everywhere you step in Alaska, western British Columbia, Washington, Oregon, western Idaho, California, or western Nevada, you stand on accreted terranes—rock brought in from the ancient Pacific and sheared off against the edge of the continent during the last several hundred million years.

Aside from terrane accretion, the most striking aspect of Farallon Plate subduction is how it pushed mountain building *eastward* through time. Even as terranes collected on the western edge of the continent, mountains blistered up inland in an east-advancing wave of volcanic fire and twisted rock. Geologists have divided this west-to-east migration into three mountain-building phases—or orogenies—the Nevadan Orogeny, Sevier Orogeny, and Laramide Orogeny. The Nevadan Orogeny is the earliest and westernmost phase, taking place during Jurassic time (202 to 142 million years ago). The Sevier Orogeny is the middle phase in time and place. It is marked by the migration of uplift eastward across Nevada and Idaho into Utah and western Wyoming throughout Cretaceous time (142 to 65 million years ago) and beyond. The Overthrust Belt of central Utah and western Wyoming represents the eastern limit of the Sevier Orogeny's fold-and-thrust belt. Beyond the Overthrust Belt, gigantic wedges of basement rock then squeezed upward to become the Foreland

---

5. *Terrane* and *terrain* are pronounced the same but have different meanings. *Terrain* describes the land surface (e.g., mountainous terrain; desert terrain). A geologic *terrane* describes a region of the earth's crust, bounded by faults, that is clearly unrelated to the rocks around it, indicating that it has moved as a unit from elsewhere. Geologic mapping in the Cordillera has identified many distinct terranes (the exact number is debated). One of the great insights of plate tectonic theory is the recognition that continents grow by the accretion of terranes at subduction zones.

~ 160 million years ago (midway through NEVADAN OROGENY)

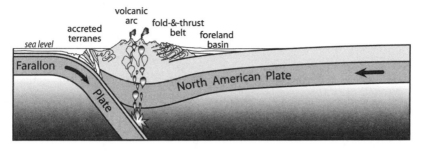

~ 100 million years ago (midway through SEVIER OROGENY)

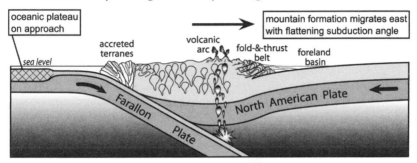

7.6 A sustained wave of mountain upheaval wracked western North America as the continent migrated west over the oceanic Farallon Plate. By early Jurassic time (about 200 million years ago), Farallon subduction had spawned an Andean-type mountain belt along the continent's then-western edge, initiating the Nevadan Orogeny. Mountain building migrated east from the Nevadan Orogeny through the Sevier Orogeny, probably in response to flattening subduction of the Farallon Plate. All the while, pieces of ocean floor were scraped off from the Farallon Plate to become accreted terranes, building the west edge of North America outward. By about 100 million years ago, an oceanic plateau was probably closing in on the Farallon Trench.

Ranges in a spectacular episode of uplift known as the Laramide Orogeny of 75 to 45 million years ago (Late Cretaceous to Eocene time).

The Nevadan Orogeny and Jurassic North America

The Farallon Plate kicked off the Nevadan Orogeny about 200 million years ago. Subduction-generated magmas blistered upward to form a volcanic arc that ran north-south roughly along today's California-Nevada border. Deep in the core of this arc, miles underground, magma congealed to form the oldest granite plutons of the Sierra Nevada Batholith.

These early plutons were just preamble, though—most Sierra Nevada granite formed later, between about 120 and 80 million years ago.

West of the arc, in the Farallon Trench, an amalgam of deep-sea sediments and slices of basaltic ocean crust gradually scraped off from the Farallon Plate. This accreted rock is on display today throughout the western Sierra Nevada foothills. Perhaps the most stunning sample is the Smartville Ophiolite—a miles-thick slab of deep ocean floor that crops out near Smartville and Grass Valley. Fractured basalt pillows—the altered remains of seabed lava eruptions—lump across its surface. They feel silky with serpentine. The rock is greenish, fibrous, and darkly luminous, like backlit stained glass. It looks alien, out of place—and it is. The Smartville Ophiolite is as far from home as a live tuna in a cornfield. It is abyssal seabed that docked on the western edge of North America about 150 million years ago, back when the continent ended in the western Sierra Nevada foothills. To see it here, more than 100 miles from the ocean and 200 miles from the nearest in-place ocean floor, is to have your mind bent with wonder at the power of the Earth's moving plates. As we'll see in chapter 13, the arrival of oceanic terranes in California is largely responsible for the gold rush.

With the addition of the Smartville Ophiolite (which arrived in a larger package of accreted oceanic rocks called the Foothills Terrane), the edge of North America fattened out to about where California's Central Valley now lies. Beyond that lay only ocean. The rest of California still lay in pieces, out in the ancient Pacific. But more soon arrived. Shortly after the Foothills Terrane/Smartville Ophiolite docked, a colossal wedge of seabed sediments and more slices of oceanic crust (ophiolites-to-be) began to dislodge from the Farallon Plate. The accretion of this material would carry forward well into Cretaceous time. Today this mishmash of accreted oceanic rock underlies California's Central Valley and crops out to view in the Coast Ranges.

While these accreting terranes were busy assembling California, masses of magma continued to roil up from the Farallon Plate, feeding the volcanic arc east of the Farallon Trench. Sideways pressure east of the arc mounted, coming from the combination of colliding terranes and the ongoing growth of the arc. The rock began to ooze east. It bent into large east-flopping folds and broke along near-horizontal thrust faults, spilling east across western Nevada. The advance of the fold-and-thrust belt bowed down the crust beyond, making a vast foreland basin that stretched east to Colorado and Wyoming. Rock debris eroded from the fold-and-thrust belt filled the foreland basin to overflowing, forming a vast riverine plain. Rivers wended their way east across this plain, pausing in lakes and swamps along the way. The result was a gorgeous mosaic of habitats—the original and authentic Jurassic Park. Stand in Wyoming

7.7 A chunk of ancient ocean floor—part of the Smartville Ophiolite—on display near Grass Valley, California, with my very good dog Scout (about two feet high) for scale. Gold-bearing quartz veins are abundant in the oceanic terranes of the western Sierra Nevada foothills—for reasons we'll explore in chapter 13. The specimen here was cut from the shaft of a nearby mine. It is serpentinized gabbro, originally from the lower reaches of the oceanic crust below the pillow basalt zone. Serpentine minerals, produced by hydrothermal alteration of oceanic crust at mid-ocean ridges, give the rock a smooth, silky feel and a banded appearance, as shown in the close-up.

or Colorado late in Jurassic time, and no mountains are in view. (The fold-and-thrust belt lies far to the west, and the Rockies don't yet exist.) Instead, you stand in an open woodland of cycads, cycadeoids, ginkgos, and primeval conifers. Crow-sized pterosaurs—flying reptiles with wings of batlike skin—swoop and spin in the air, grabbing at dragonflies. Archaic birds with long tails and beaks full of teeth watch from the branches. Turtles and crocodiles cruise in nearby pools and streams. On the bare ground (bare because grasses have not yet evolved) are footprints. Some are round and as big as hubcaps. Others have three toes, like an ostrich's, but are much larger. Big things live in these woods. An immense weight cracks a rotting branch, and the trees rustle . . . there, 60 feet up, a head like the tip of a crane munches in the coniferous treetops.

That is the foreland basin of Jurassic North America about 145 million years ago at the time of the Morrison Formation, a group of strata world-famous for its dinosaur fossils. Near Medicine Bow, Wyoming, the Morrison Formation arches up to the surface in the Como Bluff Anticline. The bones spill out of the ground in such profusion that an entrepreneur in the 1930s built a museum with walls made entirely of dinosaur bones. It still stands today as the Dinosaur Fossil Cabin Museum. (Bones were a sensible choice of construction material at Como Bluff—they were, and still are, more common than trees.) Most abundant in the Morrison Formation are the bones of humongous, long-necked herbivorous sauropods like *Apatosaurus, Diplodocus, Camarasaurus, Pleurocoelus,* and *Barosaurus*—names that roll effortlessly off the tongue of a nine-year-old but cause most adults to pause and stumble. *Stegosaurus, Camptosaurus, Laosaurus, Othnielia,* and *Dryosaurus* are there as well—herbivores, too, but not of the long-necked kind. There are carnivores at Como Bluff as well—fleet and powerful bipeds such as *Allosaurus, Ceratosaurus, Ornitholestes,* and *Coelurus.*

The Sevier Orogeny and Cretaceous North America

The onslaught of terrane accretion and mountain upheaval brought on by the Nevadan Orogeny carried on into Cretaceous time with increased intensity. The Farallon Plate plunged down beneath North America at faster rates, sending up prodigious gobs of magma. Oceanic rocks that had collected near the Farallon Trench late in Jurassic time docked to become new accreted terranes, while to the east, the volcanic arc and fold-and-thrust belt marched inland by hundreds of miles. All of this action constitutes the Sevier Orogeny.[6]

Go to Yosemite National Park or elsewhere in the Sierra Nevada, and you are surrounded mostly by Cretaceous granite. Today's Sierra Nevada range is mostly an uplifted batholith—an amalgam of thousands of plu-

6. The term "Sevier Orogeny" is sometimes limited to the growth of the fold-and-thrust belt east of the Cretaceous volcanic arc. (The name comes from exposures of the fold-and-thrust belt in Utah's Sevier Desert.) I use the term more inclusively here because it works better to view the entire Cordilleran orogenic system of trench/volcanic arc/fold-and-thrust belt/foreland basin as an integrated whole.

tons that were once magma sloshing miles below an immense Andean-style volcanic arc. If you could cut 10 or more miles of rock off the top of the Andes, you would see fresh granite akin to that in the Sierra Nevada, and glowing magma that is on its way to becoming yet more granite.

The most striking feature of the Sevier Orogeny is the eastward migration of the volcanic arc and fold-and-thrust belt. Even as magma continued to well up where the Sierra Nevada is now, new volcanic eruptions began to pop off to the east. A time-lapse movie of Cretaceous North America from outer space would look like a wave of firecrackers going off from west to east across Arizona, Nevada, and Idaho. This eastward sweep of volcanism probably reflects a decrease in the subduction angle of the Farallon Plate. A subducting plate will produce magma once it reaches about 80 to 100 vertical miles depth in the mantle. If the Farallon Plate dove under North America at a shallower angle, it would have extended farther east under the continent before reaching its magma-production depth, creating an east-advancing wave of volcanoes, plutons, and batholiths (see fig. 7.6).

Like a bow wave before a barge, the Sevier fold-and-thrust belt advanced east ahead of the volcanic arc. Throughout Cretaceous time, it pressed from eastern Nevada and central Idaho into central Utah and western Wyoming. The east-advancing front of deformed rock plowed into strata recently shed into the foreland basin, folding them over and bunching them up. The band of contorted rock eventually halted at its eastern limit about 55 million years ago, forming the Overthrust Belt.

Stretching for hundreds of miles east of the Sevier fold-and-thrust belt lay a vast foreland basin, taking up most of Wyoming, Colorado, and eastern Utah. Like its counterpart in Jurassic time, the Cretaceous foreland basin housed a rich fauna and flora. In the swamps lurked *Phobosuchus*, a 50-foot-long crocodile, stalking dinosaurs. The world's first flowers bloomed in Cretaceous time. Dinosaur herbivores thrived on these novel plants, the angiosperms (flowering plants), which would soon come to dominate the Earth's flora. Late Cretaceous dinosaurs were a different breed than those of Jurassic time. They were newly evolved species that took over ecological niches vacated by the extinction of earlier species. The wildebeests of the Late Cretaceous world were the herding bipedal hadrosaurs, so-called duckbilled dinosaurs like *Maiasaura* and *Lambeosaurus*. They migrated seasonally thousands of miles. During the long days of summer, they headed into northern Canada and Alaska to graze, trailed by skulking predators. They came back to Montana and Wyoming in the winter to breed. They dug great earthen nests for their grapefruit-sized eggs and raised their young in raucous colonies on the river floodplains of the foreland basin. Into the Late Cretaceous world lumbered *Edmontonia*, an armored dinosaur tank covered with thick plates and spikes, and the horned, rhinoceros-like *Chasmosaurus* and *Monoclonius*. Gigantic bipedal meat-shredders like *Tyrannosaurus, Albertosaurus,* and *Dromaeosaurus* hunted down these herbivores or scavenged their carcasses. They were the kings of creation, nature's most splendid horrors. Their glittering eyes commanded the world, and their breath stank of death.

The Cretaceous Interior Seaway

About 120 million years ago, the Earth's plate tectonic engine kicked into high gear and stayed there for about 40 million years. Seafloor spreading rates doubled or tripled their previous pace, cranking out new oceanic crust at rates two to three times faster than normal. Current theory holds that the accelerated pace of plate movement was triggered when several great plumes of hot mantle rock, called superplumes, rose upward from the core-mantle boundary, like an enormous heat belch from the guts of the Earth.

The consequences of this mid-Cretaceous superplume episode include volcanism and magma production on a phenomenal scale. Volcanoes erupted like mad worldwide, jacking up carbon dioxide levels in the atmosphere. (Magma contains abundant carbon dioxide along with other dissolved gasses; volcanoes spew them all into the atmosphere.) With the atmosphere soaked in carbon dioxide, the Earth entered a prolonged climatic greenhouse period—a time of exceptional global warmth. Deciduous forests like those now seen in coastal Oregon or Washington grew on Alaska's North Slope—today a tundra.[7] Dinosaurs lived there too, or at least migrated there seasonally, at latitudes where caribou thrive today.

As seafloor spreading rates picked up, the world's mid-ocean ridges swelled in height and breadth. The enlarged ridges displaced ocean water like a person does getting into a bathtub. Displaced by the tumescent ridges, the seas crept upward and flooded the low interiors of the continents. The seas invaded North America from north and south, coming up from the Gulf of Mexico and down from the Arctic. By about 100 million years ago, a large finger of the sea had advanced south from the Arctic all the way to Wyoming and Colorado, forming the vast Mowry Sea—home of the oil-rich Mowry Shale that we saw in chapter 5. By 90 million years ago, the seas had cleaved North America in two, north to south. You could have sailed on this Cretaceous Interior Seaway from the Gulf of Mexico to the Arctic Ocean, never touching land as you crossed above present Texas, Nebraska, North Dakota, Manitoba, and the Nunavut Territory.

---

7. The movement of the North American Plate has carried the continent mostly west since mid-Cretaceous time, and not north or south. Therefore, the latitude of Alaska back then was quite close to what it is today.

At its peak, the Interior Seaway stretched nearly 1,000 miles wide, from the Sevier fold-and-thrust belt in Wyoming and Colorado east to present Iowa and Missouri. The Interior Seaway would dominate the center of North America for some 20 million years, withdrawing near the end of Cretaceous time as seafloor spreading rates geared down and the world's swollen mid-ocean ridges returned to normal size.

The shoreline of the Interior Seaway formed a battlefront between the contending armies of land and sea. When the sea rose, its western shoreline advanced toward the foothills of the Sevier fold-and-thrust belt.

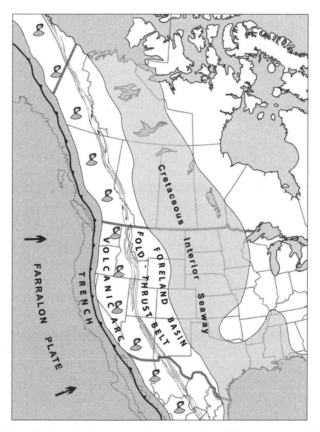

7.8 North America late in Cretaceous time, about 80 million years ago. Rising seas, pushed up by swollen, rapidly spreading mid-ocean ridges worldwide, flooded the low center of the continent to form the Cretaceous Interior Seaway. To the west lay the riverine lowlands of the Sevier foreland basin, the Sevier fold-and-thrust belt, and the Sevier volcanic arc. The Farallon Trench represents the then-western edge of the North America. The small dotted line represents today's coast. The continent in Cretaceous time did not extend as far west as it does today. It would grow to the west by further terrane accretion and by westerly stretching of the crust during the Basin and Range Orogeny, discussed in chapter 10.

When it fell, or when sediment poured east off the fold-and-thrust belt and pushed the sea back, the shoreline retreated. The back-and-forth interplay of land and sea built up a layered record of incessant environmental change. You can stand today on Upper Cretaceous outcrops in Wyoming, buffeted by desiccating winds, and pick up black coal that nearly drips with the memory of fresh water. The coal is the remains of plants that flourished in swamps along the margins of the seaway. Interleaved with the coal beds are layers of cross-bedded sandstone festooned with multiple channels, and beds of reddish-brown, root-riddled shale (once soft mud). These speak of meandering rivers and muddy, vegetated floodplains that interfingered with the coal swamps and formed the floor of the dinosaurs' world.

Moving forward in time, up through the layers, you come to beds of coarse, cross-bedded sandstone littered with the hollow molds of seashells. They tell of an advancing beach of the Interior Seaway that buried the coastal rivers and coal swamps as the sea rose. Stepping farther up through the stacked layers, you come to beds of fine sand and shale crowded with burrows and packed with shells. You can practically plunge your hand into the rock and feel things wiggle away. This shale was the soft muck of a shallow seafloor, exuberant with life, lying a few miles offshore of the beach. It covered the beach as the sea continued to rise. Forward in time and still farther up through the layers, the rock becomes fine-grained, smooth, and nearly white. This chalky limestone formed miles offshore, beyond the reach of mud washed in from the now-distant rivers to the west. Here in the open ocean, home of gamboling mosasaurs and ichthyosaurs (extinct marine reptiles with dolphin-like bodies), the only particles to settle to the seabed were the myriad remains of calcareous marine plankton. These compacted down into white limestone. Farther up-section, above the limestone, the entire sequence reverses as the sea retreats. Advance and retreat, advance and retreat; at least five major episodes of shoreline advance and retreat are recorded in the strata of the Interior Seaway.

The Foreland Ranges and the Laramide Orogeny

By late in Cretaceous time, 75 million years ago, the Cordilleran mountains ended at the still-growing Sevier fold-and-thrust belt. To the east lay no mountains, only the vast river floodplains of the Sevier foreland basin and, beyond that, the salty waters of the Interior Seaway. The mountainous geography of the West was still missing its most distinctive and magnificent family of mountains—the Rocky Mountain Foreland Ranges (fig. 7.9).

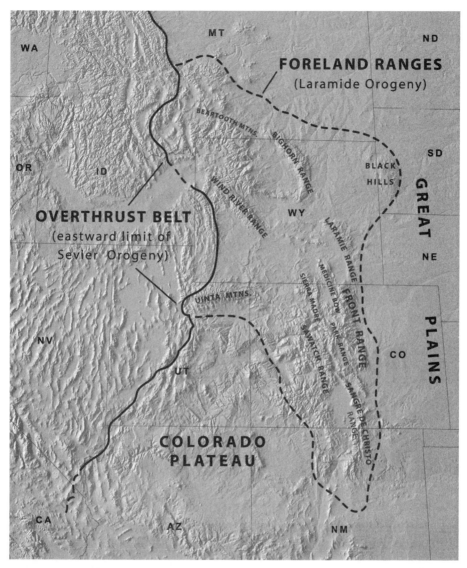

7.9 The transition from the Sevier Orogeny to the Laramide Orogeny involved an eastward jump in mountain building. The fold-and-thrust belt of the Sevier Orogeny ended at its eastern limit about 55 million years ago along a crescent-shaped line represented by today's Overthrust Belt. The Foreland Ranges rose in the foreland basin east of the Overthrust Belt. The Foreland Ranges are basement-cored uplifts—large blocks of basement rock squeezed up by sideways compression during the Laramide Orogeny.

The Foreland Ranges are the offspring of the Laramide Orogeny—the final mountain-building episode spawned by the subduction of the Farallon Plate. They rose from 75 to 45 million years ago[8] (Late Cretaceous to mid-Eocene time), overlapping with the final gasps of the Sevier Orogeny to the west. Today Colorado and Wyoming rank number one and two, respectively, in average elevation of U.S. states. But before the Laramide Orogeny, both states looked like half-submerged versions of Ohio. Both were part of the vast Sevier foreland basin, where sluggish rivers wound east to the shore of the Interior Seaway. Then, about 75 million years ago, this staid foreland basin world exploded. Great blocks of basement rock began to shoulder upward, rising like subterranean monsters out of the riverine lowlands and the floor of the sea. They pushed upward for 30 million years, shedding their overlying blankets of sandstone, shale, and limestone and lifting their fractured Precambrian backs to the sky. And geologists can only scratch their heads. How can mountains pop up within a foreland basin, 600 to 800 miles inland of a subduction trench? The standard subduction model of mountain building doesn't seem to explain it. That model, typified by the Andes, goes: (1) tectonic accretion zone, (2) volcanic arc, (3) fold-and-thrust belt, (4) foreland basin (fig. 7.4). In that model, the fold-and-thrust belt is the inland limit of mountain uplift. Yet the Foreland Ranges seem to defy this model. They reared up *within* a foreland basin, hundreds of miles inland of where mountains are supposed to be.

The puzzle of the Foreland Ranges comes into sharp focus if we compare them to the Appalachian Mountains, the older and more subdued cousin to the North American Cordillera. The Paleozoic Appalachians present in mirror image most of the major elements of the Jurassic-to-Eocene Cordillera—the accreted terranes, volcanic arc, fold-and-thrust-belt, and foreland basin. The Appalachian fold-and-thrust belt runs southwest from upper New York State through eastern Tennessee, forming the Valley and Ridge Province. To its east lie the eroded roots of a Paleozoic volcanic arc. To the west, inland, lies the Appalachian foreland basin. There are no mountains there. The Appalachian foreland basin holds only a thick stack of sediments shed west from the fold-and-thrust belt. Imagine a Laramie Range rearing up by Cleveland, or a Wind River Range soaring thousands of feet above the cornfields of Indiana. This is

8. To tell when a mountain range began to rise, you look for the oldest rock debris shed from it. As a rising mountain erodes, it sheds boulders, cobbles, and gravel into the adjacent valleys. To get the age of this debris, you look for fossils entombed in the layers, hopefully of species whose age is known from elsewhere. Even better, you might find volcanic ash beds or lava flows interleaved with the layers. The age of the ash or lava can be determined using radiometric dating.

Wyoming before the Laramide Orogeny

SOUTHWEST   NORTHEAST

Wyoming after

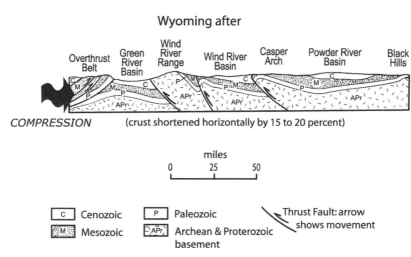

COMPRESSION   (crust shortened horizontally by 15 to 20 percent)

miles

0	25	50

	Cenozoic		Paleozoic		Thrust Fault: arrow shows movement
C		P			
M	Mesozoic	APr	Archean & Proterozoic basement		

7.10 Cross section through Wyoming showing how sideways compression from the west-southwest squeezed up the Foreland Ranges. Areas between the uplifted ranges sagged down to become basins. Reproduced by permission of the Wyoming State Geological Survey.

the oddity of the Foreland Ranges. Colorado's Front Range lies nearly 300 miles inland of the Overthrust Belt. If a parallel thing had happened during the uplift of the Appalachians, Chicago would wake up each morning to mountain views like those seen from Denver.

To understand the Foreland Ranges, we need to consider the mechanism that squeezed them up. The faults and folded rocks of the ranges show that they arose by massive sideways compression of the deep continental basement, coming from the west-southwest. Squeeze a baseball-sized wad of clay sideways between your hands. The clay rises up and out, escaping the pressure. That roughly mimics the uplift of Utah's Uinta Mountains or Colorado's Front Range—basement-cored uplifts bounded on both sides by faults (the surfaces of your two hands). Other ranges, such as the Wind River Range and the Laramie Range, squeezed up along single faults. Of all the Foreland Ranges, the Wind River uplift is the most staggering. The highest Wind River peaks nearly reach 14,000 feet. The same basement rock of those peaks lies 30,000 feet *below sea level* under the Green River Basin, on the other side of the Wind River Thrust Fault. The difference—44,000 feet, or more than eight miles—is the minimum distance that the rock of the Wind River Range has risen from its origins in the basement. Imagine a fault splitting your neighborhood in two and lifting one side 44,000 feet up: 15,000 feet higher than Mount Everest, 9,000 feet above where passenger jets usually fly. This is what happened to lift the Wind River Range, although not all at once. It took 30 million years and many thousands of earthquakes to accumulate this uplift.

The key to understanding the Foreland Ranges is to figure out how the Farallon Plate could have compressed the continental basement so far inland of the Farallon Trench. Currently, the most widely accepted theory is flat subduction. The idea is that the Farallon Plate, for a time, slid nearly horizontally underneath the western United States between Montana and New Mexico, like a board sliding flat beneath a carpet.

Volcanic evidence generally fits with the flat subduction theory. As the Foreland Ranges were rising in Wyoming and Colorado, there was a striking lull in volcanic activity directly to the west, throughout Nevada and Utah. Volcanoes erupted with abandon in these areas both before and after the Laramide Orogeny, presumably because the Farallon Plate was subducting at a more normal angle and generating magma in typical fashion. Flat subduction during Laramide time may explain the volcanic lull. A nearly horizontal plate would not reach deep enough into the mantle to generate magma. In addition, from about 80 to 55 million years ago (nearly coincident with the uplift of the Foreland Ranges), volcanoes blasted off in an east-migrating wave from Oregon across Idaho and from southern California across Arizona. The two trends together—the volcanic lull in Nevada-Utah and the eastward migration of volcanism to the north and south of there—make sense with flattening subduction.[9]

As sensible as it may seem, the flat subduction theory for the Laramide Orogeny wouldn't be worth the papers that have been written on it if it can't be shown that flat subduction can actually happen. True to the legacy of Charles Lyell, geologists are happier with explanations for events in the past if they can see the Earth doing similar things today. We began this chapter using the modern Andes as a model for the ancient Cordillera. That model worked fairly well for the Nevadan and the Sevier orogenies. But the Foreland Ranges left us waving our arms about flat subduction without knowing whether the Earth can actually *do* such a thing. As it turns out, though, the Andes have more to teach us.

9. North into Canada and south into Mexico, the Farallon Plate seems to have subducted at a normal angle during the Laramide Orogeny. Mountain building in those regions was business as usual, with a classic Andean-type volcanic arc and fold-and-thrust belt, and no wacky basement-cored uplifts popping up in the foreland basin.

~ 100 million years ago (midway through SEVIER OROGENY)

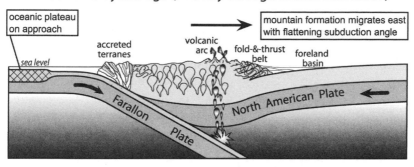

~ 60 million years ago (midway through LARAMIDE OROGENY)

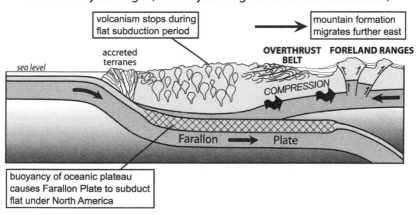

7.11 Subduction of an oceanic plateau on the Farallon Plate may have caused the plate to subduct flat beneath North America, compressing the continental basement to squeeze up the Foreland Ranges. A similar process is happening below a portion of the Andes today, where flat subduction of the Nazca Plate is pushing up the Pampean Ranges (see figure 7.5).

By tracking seismic waves through the Earth's interior, we can take a coarse picture, like an X-ray, of the planet's innards. The technique has recently allowed us to "see" hundreds of miles into the mantle below the Andes. Remarkably, this seismic view shows that between about 28 and 33 degrees south latitude (a 350-mile-wide swath roughly between Copiapó and Valparaiso, Chile), the Nazca Plate appears to be sliding east nearly flat beneath the Andes. The consequences are reminiscent of the Laramide Orogeny, albeit on a smaller scale. For instance, there is no volcanic activity in the Andes here, presumably because the flat Nazca Plate lies above its magma-generation depth. Moreover, modest basement-cored uplifts — miniature versions of the Wind River Range or the Front Range—are squeezing up far inland, in the foreland basin of Argentina, 400 miles east of the Peru-Chile Trench (see fig. 7.5). These mountains, called the Pampean Ranges, appear to be wedging upward in response to sideways compression as the Nazca Plate scrapes flat under South America. The easternmost Pampean uplift, the Sierra de Córdoba, rises several hundred miles east of the Andean fold-and-thrust belt. The city of Córdoba nestles against its eastern flanks, a geological sister city to Denver at the foot of the Front Range.

Significantly, this flat subduction region below the Andes occurs where the Juan Fernández Rise—a particularly thick portion of the Nazca Plate—is heading into the Peru-Chile Trench. This can't be coincidence. The thickened oceanic crust of the Juan Fernández Rise is probably causing the flat subduction. Since oceanic crust is less dense than mantle rock, the thicker the oceanic crust that caps an oceanic plate, the less dense overall is the plate.[10] The Andes teach us that if you subduct a plate capped by a particularly thick section of oceanic crust, the extra buoyancy can float the subducted plate upward in the mantle so that it scrapes flat beneath the continental plate overhead. This is our best guess for what happened during the Laramide Orogeny. We think that an oceanic plateau,[11] riding on the Farallon Plate, dove under western North American to hoist the Foreland Ranges (fig. 7.11).

SUBDUCTION OF the Farallon Plate under North America built the mountainous American West. Instead of creating a narrow mountain belt hugging North America's west coast—like most of the Andes—Farallon Plate subduction pushed up mountains far inland, particularly during the flattening (Sevier Orogeny) and flat (Laramide Orogeny) subduction phases. This meant that westbound emigrants entered mountainous country while still 1,300 trail miles from California. Their experiences on that route were determined, in no small part, by our continent's own multimillion-year history of westward migration. We rejoin those folks now as they take on the harder half of their road west.

10. Ocean floor basalt is about 10 percent less dense than peridotite—the main rock of the upper mantle. (Basalt density is about 2.9 to 3.0 grams/cc; peridotite is about 3.3 grams/cc.) Therefore, the more of a subducting plate that is made of basalt, the relatively more buoyant it becomes in the mantle.

11. There are several dozen oceanic plateaus scattered across the world's ocean basins. They form vast welts of volcanic basalt, hundreds of miles across, that rise from the abyssal ocean floor. While typical oceanic crust is three to four miles thick, the basalt lava beds of oceanic plateaus may stack up 20 miles thick. The world's largest is the Ontong-Java Plateau east of New Guinea. Two times larger than Texas, it contains enough basalt to cover the 48 coterminous states more than two miles deep.

✳ 8 ✳

MOST GODFORSAKEN COUNTRY

South Pass, 7,550 feet high, the backbone of the continent, where Atlantic and Pacific waters part. One thousand miles of overland trail stretched east, back to the Missouri. One thousand more lay ahead. And so, with the parties over and the flutes, fiddles, and whiskey packed away into the wagons, the emigrants bid farewell to the Atlantic side of the continent, possibly forever. Sad and lonely reflections often welled up at this moment. Alonzo Delano stood at the pass on June 29, 1849, and took "a parting look at the Atlantic waters, which flowed towards all I held most dear on earth."

> As I turned my eye eastward, home, wife, and children, rushed to my mind with uncontrolled feeling, and in the full yearnings of my heart, I involuntarily stretched out my arms as if I would clasp them to my bosom; but no answering look of affection, no fond embrace met me in return, as I was wont to see at home, but in its place there lay extended before me barren reaches of table land, the bare hills, and desert plains of the Sweet Water, while long trains of wagons, with their white covers, were turning the last curve of the dividing ridge, their way-worn occupants bidding a long, perhaps a last adieu, to eastern associations, to mingle in new scenes on the Pacific coast.

Up to South Pass, the emigrants had had a relatively easy time rolling west along the valleys of east-flowing rivers—the Platte, the North Platte, and the Sweetwater. West of South Pass, the landscape becomes less cooperative. Mountains, rivers, and canyons tend to run north-south,

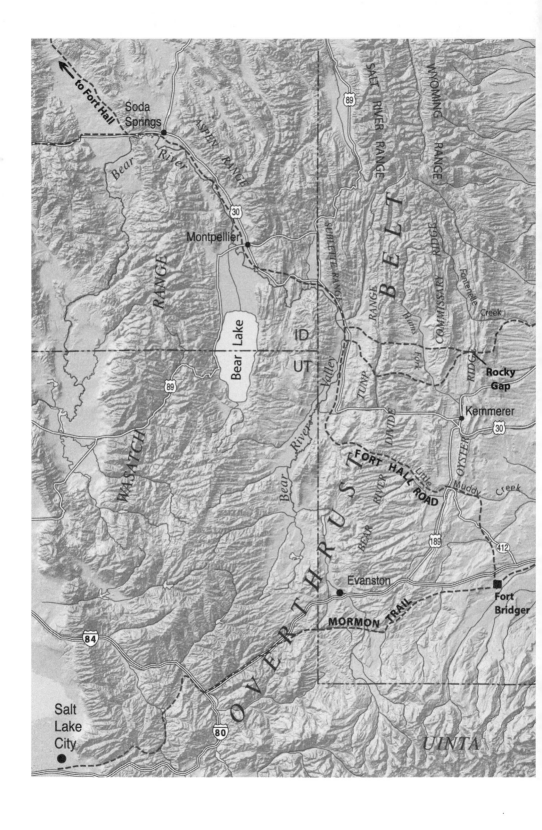

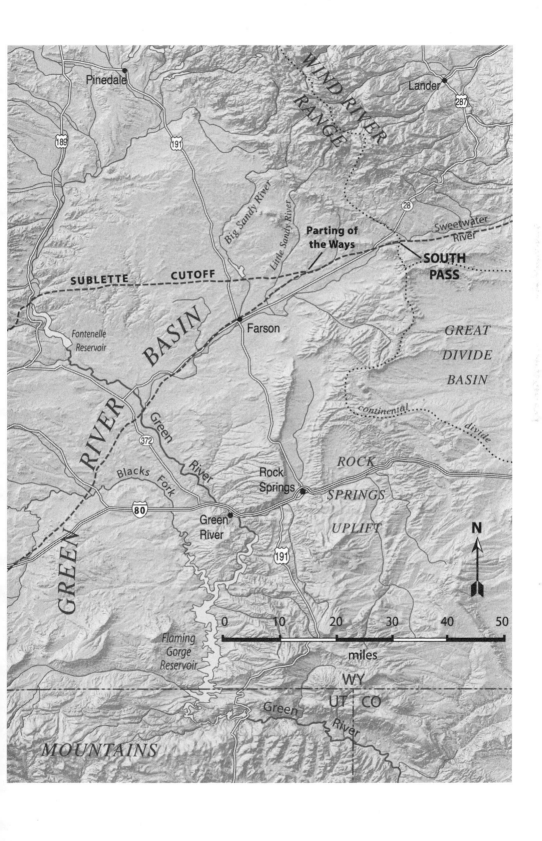

8.2 J. Goldsborough Bruff's first sketch from the Pacific side of the continent, drawn on August 3, 1849. The view looks east, back to South Pass, which Bruff labels "S.P." The buffalo hunt he portrays would be just about the last one of the journey, for buffalo did not venture far west of South Pass. Reproduced by permission of the Huntington Library, San Marino, California.

8.1 (*previous pages*) From South Pass, emigrants had several options for reaching California. They could go to Salt Lake City via the trail to Fort Bridger, and then take either the Salt Lake Cutoff or the Hastings Cutoff west from there. Or they could turn north from Fort Bridger onto the Fort Hall Road (part of the original Oregon Trail), which led to Fort Hall on the Snake River. Or they could bypass Fort Bridger entirely by taking the Sublette Cutoff, which connected up with the Fort Hall Road in the Bear River Valley. Whichever way they went, they had to cross the vast, bleak Green River and the rough terrain of the Overthrust Belt.

cutting across westerly routes of travel. From South Pass to the Humboldt River—more than 500 trail miles—there are no direct westerly river corridors except for one brief stretch along the Snake River. A look at a map of the emigrant trails reveals this shift in landscape. Up to South Pass, nearly everyone followed one road. Beyond the pass, the trail system

sprouted cutoffs like a branching bush as emigrants tried to find faster or easier routes across the corrugated land.

Heading west from South Pass, the trains rolled down into the sprawling Green River Basin. "My companions denounce this as the most Godforsaken country they ever saw," A. J. McCall wrote in July 1849. "Destitute of timber, of grass, of game — even water scarce. They cannot see for what purpose it was raised from the sea. They forget that it is not for us to question the purposes of an all-wise Creator."

The Green River Basin forms a vast, windswept plain that takes up most of southwestern Wyoming. As the Foreland Ranges rose during the Laramide Orogeny, they broke up the vast Sevier foreland basin into a dozen or more smaller basins. Sagging down between the rising Wind River and Uinta ranges, the Green River Basin become a receptacle for rock debris sloughed off the rising mountains. The debris piled up thousands of feet thick, rising higher and higher like a flooding tide, eventually cresting at the level of today's subsummit surface. Then, with the onset of the Exhumation about 5 million years ago, the Green River and its tributaries began to dig out the basin. Take the current elevation of the land surface in the Green River Basin (7,700 feet in the north, sloping down to 6,100 feet in the south) and subtract that from the level of the subsummit surface on the mountain flanks (about 10,000 feet), and you know roughly how much gravel, sand, silt, and clay has been scooped out of the basin since the Exhumation began. (Enough, as it turns out, to cover all of Wyoming 200 feet deep.) Most of it went south with the Green River, but much of the smaller debris—the silt and clay—probably headed east with the wind to settle as dust on the Great Plains.

Crossing the Green River Basin redefines monotony. The plains roll on endlessly, blanketed by the same wearisome mantle of sagebrush and greasewood. Outside of the river bottoms and the few towns, there is not a tree in sight. Scabby buttes, eroded from the stacked limestone and shale layers of an ancient lakebed, pop up here and there across the plains. The landscape is riven with ravines, most of them bone-dry in summer. The scenery has hardly changed since emigrant days. For Edwin Bryant, it was "scarcely possible to conceive a scene of more forbidding dreariness and desolation than was presented to our view on all sides." Only the wind seems happy in the Green River Basin. It shrieks with glee across the plains, sweeping up wraithlike clouds of grit. "It has been windy, and there is nothing but sand—sand all around us, which is drifting constantly, filling our eyes and ears, as well as the frying pan," A. J. McCall groused, adding, "It is not strange that it affects the temper of the men—marring all good fellowship." After a day of pummeling by Wyoming's biggest bully, I can vouch that nothing is more welcome than a building—*shelter!*—even if it is a run-down gas station in a run-down town like Farson, a forlorn little hamlet marooned in the sagebrush wilderness of the Green River Basin. I sipped burnt coffee there one afternoon, hiding from the wind, while leaves, newspapers, and other flotsam flew past the windows. Was this typical? Oh yes, the attendant sighed. The windows appeared cloudy. A closer look showed that they had been etched by windborne sand. Merciless wind and winter beat up the small rural towns of Wyoming. The results are evident as potholes, peeling paint, broken roofs, leaking pipes, and plywood windows. Yards spill over with rusted cars, wrecked parts, writhing heaps of hose and pipe, and tires—many, many tires. Anything that might be useful, might a save a few dollars one day, joins the heap. But local folk brim with friendship and conversation, a pleasant upshot of life with so much open space and so few people to fill it.

THE RIVERS of the Green River Basin drain south. The drainage system looks like the veins of a leaf, with the Green River as the central vein. The smaller veins—the tributaries—flow southeast or southwest to join the

Green, which exits the basin south through the Uinta Mountains via Lo-dore Canyon—a superposed chasm sawed through the mountains during the Exhumation. For westbound emigrants, the south-flowing drainage system presented two options for crossing the Green River Basin: a long way—the trail to Fort Bridger—and a hard way—the Sublette Cutoff.

The trail to Fort Bridger (the original Oregon Trail) generally followed the riverine veins. It headed southwest along the Big Sandy River to the fort, and then turned northwest toward Fort Hall on the Snake River. After the Mormons settled the Salt Lake Valley in 1847, many emigrants chose to head to California by way of Salt Lake City. To do that, they followed the trail to Fort Bridger, and then, instead of turning northwest, they kept on southwest through the Wasatch Range down to the Mormon city. From there, either the Salt Lake Cutoff or the Hastings Cutoff connected them with the Humboldt River—the final leg of the road to California.

If you didn't plan to go through Salt Lake City, and you didn't need to resupply at Fort Bridger, the Sublette Cutoff was the shortest route. The cutoff headed straight west across the Green River Basin and across the eastern ridges of the Overthrust Belt to the Bear River Valley. There it joined the trail coming up from Fort Bridger. The cutoff saved about 50 miles and several days of travel compared to going by way of Fort Bridger, but at a price. Right at the start there was a stretch of 45 miles without water. Also, because the cutoff headed due west across the south-flowing drainages, it was rougher than the Fort Bridger trail.

The trail fork for the Sublette Cutoff comes up about 20 miles west of South Pass. Ten miles farther, the cutoff intersects the Big Sandy River—the last water until the Green River 45 miles to the west. At ten feet wide and one foot deep, the Big Sandy is big only in comparison to the Little Sandy, which lies just to the east. But no matter—for the emigrants, it held water enough. Here they loosed their stock to graze, drink, and rest. Meanwhile, they filled every available container with water and stuffed the wagons with cut grass for feed on the drive ahead. It was a wretched spot to get ready for the 45-mile dry run, "a nasty, dirty place, where the many who have camped here before have left all their filth and offal."

Most set out for the Green River in the late afternoon, traveling all night to reduce the sun's torment and to conserve water. As Alonzo De-lano explained, "The desert over which we were to pass was an arid plain, without a drop of water, or a blade of grass, the soil being of soft, dry, ashy consistence. The dust was an impalpable powder, and the dense clouds which arose almost produced suffocation." The trail was level for the first leg of the journey. Approaching the dissected tablelands border-ing the Green River, the route roughened, dropping down into ravines

8.3 During the Exhumation, the Green River sliced down through thousands of feet of rock strata, leaving high bluffs along the flanks of its valley. The upper photo looks east across the Green River. In the distance are 400-foot-high bluffs that emigrants on the Sublette Cutoff had to descend to reach the river. The lower image shows J. Goldsborough Bruff's sketch of this descent on August 5, 1849. In their desperation to reach the river after 45 waterless miles, many emigrants took this descent too quickly, wrecking their wagons and teams. The area is a few miles south of present La Barge, Wyoming. Reproduced by permission of the Huntington Library, San Marino, California.

and rising up over intervening ridges. "Such a rough and barren country is enough to kill the Devil, much more the cattle."

Two full days (or two nights and an intervening day) of waterless travel passed before dust-red eyes spied the verdant band of cottonwoods lining the Green River. But precipitous bluffs, 400 feet high, divided the weary pioneers from salvation. "The hill descending to the Green River is the worst we have yet come across being very long steep crooked & rocky," Franklin Starr wrote. Lured by the siren river so near, many emigrants were not careful enough on this final steep descent. The result was wreckage, as J. Goldsborough Bruff saw on August 5, 1849. "From the crest [of the bluffs], down to the base, right and left, were fragments of disasters, in shape of upset wagons, wheels, axles, running-geer, sides, bottoms, &c. &c. Nothing daunted," Bruff continued, "we double-locked, and each teamster held firmly to the bridle of his lead mules, and led down, in succession, till the whole train reached the valley below, about 1/3 of a mile, without accident."

Freed of their traces, the thirsty oxen and mules "rushed pell-mell down into the river," William Swain recalled. "It was a beautiful, clear stream," he added, "and they stood in it drinking and cooling their feet for a long time." Thus ended the dry run of the Sublette Cutoff. It had saved 50 miles, but that the price of "numerous dead oxen and wrecks of many wagons."

WHETHER THEY crossed the Green River Basin by the Sublette Cutoff or the trail to Fort Bridger, the Green River—several hundred feet wide and dangerously swift—lay as a barrier across the emigrants' path. The Green would be the last large river that California-bound emigrants would have to cross. (Those headed for Oregon would still have to contend with crossing the Snake River.) The challenge of crossing the Green varied with the time of year and the annual snowpack in the Wind River Range. Emigrants arriving at the river in August, after much of the winter snow had melted, sometimes found it low enough to ford. Most arrived earlier, though, in late June or early July, and saw it as Margaret Frink did—running "high, deep, swift, blue, and cold as ice." At such high water, a ferry was the only safe way to cross. By 1847 several ferrying operations, run by mountain men or Mormons from Salt Lake City, lay scattered up and down the Green River at the common crossing points. Emigrants forked over tolls ranging from $3 to $16 per wagon (roughly $60 to $320 in today's dollars), depending on demand and river level. "While others are chasing wealth, they are catching it," John Edwin Banks ruefully noted of these ferrymen. Livestock were generally swum across the river, while wagons and people were ferried. "A rope with pulleys was stretched

8.4 Reconstruction of a Green River ferry at Lombard Crossing—one of the common crossing points of the Green River on the trail from South Pass to Fort Bridger.

across the river, and the current carried the boat across," recalled John B. Hill. "When we were nearly across, the upper edge of the boat dipped . . . and I thought we would be swamped instantly . . . and drown to the last one of us. At the time, the Green River was booming."

During the peak emigration years, wagons often backed up for several days along the east bank of the Green, waiting for the ferries. "Still in camp waiting to cross," wrote an exasperated Amelia Stewart Knight on June 28, 1853. "Nothing for the stock to eat," she added. "As far as the eye can reach it is nothing but a sandy desert, and the stench is awful." She was still waiting the next day. "The wagons are all crowded up to the ferry waiting with impatience to cross. There are 30 or more to cross before us. Have to cross one at a time. Have to pay 8 dollars for a wagon; 1 dollar for a horse or cow. We swim all our stock." The only benefit of the wait was the shade of the towering cottonwood trees that lined the banks of the Green. These were the first tall, shady trees the emigrants had seen since the North Platte River, now several weeks and several hundred miles behind. For Edwin Bryant, the trees afforded "an agreeable and picturesque contrast to the brown scenery of hill and plain on either side."

The tributaries of the northern Green River Basin boasted some of the greatest beaver waters in the west. By the early 1800s, the basin had become the heart of a booming Rocky Mountain fur trade. Mountain men wandered the dendritic streams, searching out the best beaver waters. Many took one or several Indian wives, settled down, and called the Green River Basin home. Forty-niner Oliver Goldsmith described one group he met. "They had adopted the native custom of taking as many wives as they could feed. Each man had from three to six wives. . . . They were all crazy for whiskey and offered us any one of their wives for a drink." The most celebrated of the Green River Basin fur trappers was Jim Bridger. A consummate mountain man, Bridger was both a successful trapper and a superlative explorer who held detailed maps of thousands of square miles of the Rockies in his head. He was the first white man to set eyes on Great Salt Lake (which he believed at the time to be an arm of the Pacific Ocean), and the first to report on the geysers and petrified fossil forests of Yellowstone. "He was a born topographer; the whole West was mapped out in his mind," recalled General Grenville Dodge, who hired Bridger in 1865 to help scout the route of the transcontinental railroad through the Rockies.

As the fur trade declined through the 1830s, Jim Bridger thought about new ways of making money. He saw his opportunity in the newly blazed Oregon Trail. The emigrants, Bridger decided, "are generally well supplied with money, but by the time they get there [to the Green River Basin region] are in want of all kinds of supplies." Keen on intercepting emigrant dollars, Bridger and a partner built a trading post in 1843 on the Black Fork of the Green River. Fort Bridger was not a fort in the military sense, only "two or three miserable log-cabins, rudely constructed, and bearing but a faint resemblance to habitable houses," according to Edwin Bryant. But Bridger made up for the deficiencies of his establishment with his warmth, humor, and sage advice about trails and terrain. "[Bridger] received me most cordially and replied freely to all my inquiries," A. J. McCall remembered. "He drew for me a map in the rude floor with charcoal, of the great basin, marking the course of the streams and mountain passes, showing clearly that he had the whole region mapped in his mind. . . . I listened to him and his fellow mountaineers with great interest for several hours, discussing matters pertaining to their craft and the perils of their adventurous lives." Bridger freighted in his goods from Fort Laramie on the western Great Plains, and after 1847 probably supplemented his supplies from the Mormon settlements at Great Salt Lake. By 1849 he was well positioned to reap a handsome profit from gold rush emigrants. Richard Ackley reported prices at Fort Bridger for "sugar and coffee $1.00 per pound, butter 75c, potatoes $10.00, and flour $12.00 per hundred, eggs 60c per dozen and other things in proportion."

THROUGHOUT THE Green River Basin, many "elevated ~~buttes~~ of singular configuration" rise from the rolling plains, sticking up like flat-topped warts on a sagebrush skin. They stand several hundred feet high, with eroded edges that descend in irregular steps. A closer look reveals that the jutting ledges are made of layers of beige sandstone and off-white limestone, while the soft slopes are comprised of tan, green, and rusty-brown mudstone. To Edwin Bryant, the buttes looked like islands. "The plain appears at some geologic era to have been submerged, with the exception of these buttes, which then were islands, overlooking the vast expanse of water." Bryant was wrong about the islands, but right about the water. The Green River Basin was once the site of an immense lake. The buttes are the eroded remains of once-continuous sedimentary layers that blanketed the lake's floor.

Ancient Lake Gosiute, nearly twice the size of Lake Erie, filled the Green River Basin early in Eocene time, about 50 million years ago. It formed as the Foreland Ranges were completing their rise to block out the current borders of the basin. Lake Gosiute had a twin of roughly equal size, Lake Uinta, on the south side of the Uinta Mountains in Utah. The two lakes grew and shrank with the vagaries of climate. At its highest levels, Lake Gosiute overflowed into Lake Uinta by a connection around the east end of the Uinta Mountains. About midway through its existence, Lake Gosiute shrank and grew murky with chemical precipitates, particularly trona. Commonly called soda ash in its processed form, trona is a hydrous form of sodium carbonate used in manufacturing glass, ceramics, paper, steel, detergents, and baking soda. Astonishing amounts of trona precipitated in Lake Gosiute. The Green River Basin holds the world's largest trona deposit—more than 130 billion tons—making Wyoming the world's leading supplier of commercial soda ash. Most of it comes out of deep mines near the town of Green River. Miners use 10-foot chain saws and dynamite to excavate the soft trona from mazelike warrens 1,000 or more feet underground. All told, Lake Gosiute's layered beds of trona, shale, and limestone stack up nearly a half-mile thick and span 8 million years.

The fossils of Eocene Lake Gosiute reveal a world apart. Crocodiles and turtles lurked where antelope and elk today paw snowdrifts for winter forage. Palm trees swayed and dripped in moist tropical breezes where today only xeric sagebrush and greasewood thrive. Travel back in time to this 50-million-year-old lakeshore, and the air caresses you with soft humidity. It shimmers with heat and carries the gentle funk of shoreline swamps where old life is disassembling into nutrients and soil from which new life will sprout. Cattails, horsetails, and ferns cluster in the shade of broad palm tree crowns. Back from the shore, on the floodplains of the rivers that feed the lake, stand dense groves of poplar, willow, wing nut, sycamore, alder, and elm. Cat-sized horses, with four toes on dainty feet, creep through the underbrush. The warm air hums with insects—flies, dragonflies, mosquitoes, and bees. A sparkle sweeps the water surface, as if from a handful of tossed pebbles. It is a school of *Knightia*—finger-sized herring that swarm by the millions in the lake. They are chased, perhaps, by a perch or bowfin or six-foot gar. Frigate birds swoop overhead with an eye to robbing a meal from one of the long-legged shorebirds that probe the muddy shallows.

Lake Gosiute's tropical visage mirrored that of the early Eocene Earth. Fifty million years ago, during a time interval called the Eocene Thermal

8.5 Fossils from the Eocene Green River Formation, the remains of Lake Gosiute. The large fish on the upper slab, *Diplomystus dentatus*, died in the act of swallowing the smaller *Knightia eocaena*. The turtle is *Trionyx sp.* (Photographed from exhibits at Fossil Butte National Monument in southwestern Wyoming.)

Maximum, the Earth was experiencing one of the greatest episodes of global warming in its history. Palm trees grew in Alaska. A Congo-like tropical jungle cloaked the British Isles. Tortoises and alligators—animals that cannot endure cold winters—lived on now-frigid Ellesmere Island, 800 miles north of the Arctic Circle.[1] With all the talk today about global warming, anyone could be forgiven for thinking that it is something unusual. It's not. The Earth has swung through wild climatic change all through its history. Past climates at times have been much warmer or colder than today. Human society (an afterthought in the scope of geologic time) has not been on the planet long enough to see radical climate change—yet. To see where today's global warming may be taking the Earth and its passengers, we can look to the Eocene Thermal Maximum—to the world of Lake Gosiute. The view isn't pretty. In a world so warm that alligators colonize lands where polar bears now live, polar bears and countless other cold-adapted life-forms would be pushed off the planet forever. Global warming on such a scale would throw human society into unimaginable chaos. Shifting rainfall belts might lead to the collapse of agriculture, with consequent worldwide famines. The warmed oceans would likely hurl intensified hurricanes at coastlines already flooding from the melting polar ice caps. As the seas invaded the coastal plains of the world, teeming millions of dispossessed people would stream inland, creating a refugee crisis of nightmarish proportions. Miami, New York, and Los Angeles would be at shoal water depths, awash with waves, with algae swaying from the rooftops and fish swimming the streets.

LEAVING THE remains of Lake Gosiute behind, the emigrants ascended gullied plains toward the Overthrust Belt, looming dark on the western horizon. The Overthrust Belt is a fold-and-thrust belt—a heap of bent and broken rock shoved miles from the west during the Sevier Orogeny. From above, the serial north-south ridges and valleys stand out like the folds in a carpet pushed against a wall. The easternmost ridge—the first one that the emigrants came to—is called Oyster Ridge. It rises from the sagebrush plains of the Green River Basin like a wall, several hundred feet high, running north-south as far as the eye can see. The sedimentary layers that make up the ridge—mostly shales, sandstones, and coal beds of the Cretaceous Frontier Formation—tilt steeply west. The uppermost layer, etching out a sharp, white ridgeline, is a thick bed of Cretaceous

1. Knowing that land masses drift about on the Earth's moving tectonic plates, you could fairly ask whether Alaska, the British Isles, or Ellesmere Island were as far north 50 million years ago as they are today. As it turns out, their latitudes have not shifted appreciably since that time. We can tell because the angle at which the Earth's magnetic field intersects the planet's surface changes with latitude. By measuring this angle in rocks of different ages, we can tell how much a region has drifted north or south over time.

8.6 View northwest toward Oyster Ridge—the beginning of the Overthrust Belt in southwestern Wyoming. A thick bed of Cretaceous oysters forms the white ridgeline. The emigrant trail from Fort Bridger passes through the ridge via a superposed gap (not visible here) cut by Muddy Creek. Wyoming Highway 412 in the foreground takes the same route today.

oysters, with shells as large as a man's hand or larger. The oyster bed covers and protects the west face of the ridge like a sloping roof. During the Exhumation, the surrounding landscape headed south in tiny pieces down the Green River's tributaries, leaving Oyster Ridge—with its resistant oyster cap—standing high. East-flowing streams superposed themselves onto the ridge, slicing downward to cut gaps that grant the only passages west. Emigrants on the trail from Fort Bridger passed through Oyster Ridge via Muddy Gap, while those on the Sublette Cutoff passed through Rocky Gap about 30 miles to the north. "The bluff is an immense oyster bed," James Bennett noted at Rocky Gap, "the whole of the summit being covered with fossils of this description of a large size; some of them measuring nine inches in length."

If you could somehow slide all the rocks of the Overthrust Belt back to their starting points and shake out the wrinkles, you would find that some of them once lay as much as 100 miles to the west. The dispossessed rocks bent into gross contortions as they were pushed east. To gaze on these rocks today, shoved to their present location from beyond the visible horizon and folded up like wet laundry, is to stand humbled before the Earth's quiet power. The rocks migrated east along five great thrust faults and at least twice as many smaller ones. The faults are like gentle ramps that start miles below ground to the west and slope gently up to the east. From west to east, the five big thrust faults are the Paris-Willard Thrust, Meade-Laketown Thrust, Crawford Thrust, Absaroka Thrust, and Hogsback Thrust. The westernmost thrusts were the first to move and the first to stop. The rocks slid east and crumpled up along one fault until the accumulating pile became too thick and resistant. Then a new fault would break out east of the old one, then another one east of that, and so on, stacking the thrust sheets up like shingles. (Fig. 8.7 shows the sequence.) In some places in the Overthrust Belt, single oil-well drill holes have punched through as many five stacked thrust sheets. The faults favored weak zones in the rock, often breaking out along beds of soft shale, gypsum, or salt. Individual thrust faults merge downward, like branches on a tree trunk, into a single, nearly flat fault, thousands of feet down, called a décollement—a master fault that divides rocks that have slid east (above) from those that have stayed put (below).

Twenty-five miles of ridges and valleys divide Oyster Ridge from the Bear River Valley. These miles encompass some of the highest country on the Oregon-California Trail. The valleys lie between 6,500 and 7,500 feet above the sea, while the ridgelines, rising to the north, crest in the snowcapped 10,000-foot peaks of the Wyoming Range and Salt River Range. Dozens of shining clear-water creeks, fed by snowmelt, run along the valley floors. Their chill waters pause here and there in little pools teeming with trout, and then hasten onward, chattering and tumbling between willow-lined banks. The valleys and ridge slopes are painted with pastels of beige grass, gray-green sagebrush, and yellowing aspen, while green pines cluster on the ridgetops. For the emigrants, having just come through the dreary ordeal of the Green River Basin, the mountain splendor of the Overthrust Belt was pure enchantment.

"We were delighted to be among trees once more," Margaret Frink wrote from one of the wooded ridgetops. William Swain was likewise cheered by the "verdure of the country [that] begins to assume the look of Spring. Some grass green, vegetation generally fresh, and flowers in bloom. . . . [W]e found good grass, green and luxuriant, which we have not found since we left Deer Creek on the Platte." Emigrants traveling the Sublette Cutoff often camped in the gorgeous valley of Fontenelle Creek. "After having encamped for so long a time in the barrens and in the sand," Israel Hale explained, "we find it quite a luxury to encamp once more on the green grass and more especially in as handsome a valley and by a fine a creek as this." Augustus Burbank celebrated Independence Day 1849 at Fontenelle Creek. "We are entertained this evening with various kinds of music. On both sides of the stream, drum, fife, key

Before the Sevier Orogeny
(Jurassic)

WEST
(Nevada / Idaho)

EAST
(Utah / Wyoming)

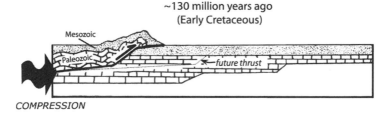

~130 million years ago
(Early Cretaceous)

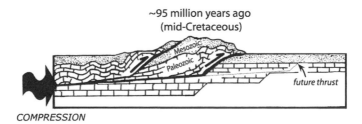

~95 million years ago
(mid-Cretaceous)

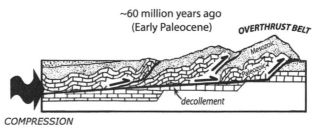

~60 million years ago
(Early Paleocene)

8.7 The Overthrust Belt represents the eastern limit of a fold-and-thrust belt pushed eastward during the Sevier Orogeny (chapter 7). Sideways compression from the west shoved rocks east along large, low-angle thrust faults. The oldest faults broke out to the west, and as the rocks slid east and piled up, newer faults broke out to the east. The result is a wrinkled, mountainous landscape of folded and faulted rocks, some of which have traveled more than 100 miles from their points of origin. Reproduced by permission of the University of Utah Press.

bugle, etc. beside various voices talking, laughing, halloing, and crowing. From the noise and bustle, one would fancy himself to be on the levee at St. Louis." J. Goldsborough Bruff found some novel game along Fontenelle Creek. "Innumerable large black mice here . . . fat and very soft & silky: not at all shy, probably unacquainted yet with man's destructive propensity, of which I, however, convinced one by knocking it over, roasting & eating it.—Found it very tender & sweet."

Mountain beauty and tasty mice notwithstanding, there were still 25 miles of roller-coaster ridges and valleys to cross to get to the Bear River Valley. It was hard, panting work—as J. Goldsborough Bruff described on August 10, 1849:

> The company had some very rough traveling over high and very rocky ridges, and 2 narrow winding & very dangerous descents. . . . Later in the afternoon we moved on again, soon ascending a succession of ridges,—higher and higher, to the greatest elevation of the divide;—the descent from which was a series of precipitous places:—deep sand, with loose stones: and dangerous.

Oxen, already worn from the strain of crossing the Green River Basin, died by the dozens on the rough ridges of the Overthrust Belt. "Dead animals all the way up, the stench intolerable," Byron McKinstry complained in July 1850. He continued:

> We have had the road strewed with putrid carcasses ever since we left the Platte. As soon as an ox dies, he bloats as full as the skin will hold (and sometimes bursts), and his legs stick straight out and soon smells horribly. . . . When they are nearly decayed I think there is frequently three or four bushels of maggots about the carcass. At the top of the steepest pitch this morning lay eleven dead oxen. They pulled up the pitch and died when they stopped to rest. . . . Thus they lie strewed on every hill and in every valley, thus poisoning the otherwise pure air. The most die after getting over some hard place, or long stretch.

The ridge ascents were tough, but the precipitous descents were more dangerous. John Edwin Banks professed himself "astonished how men ever found a passage through these rugged piles of rock and earth." James Pritchard's company was "compelled to let our wagons down in part by attaching ropes and letting the men hold on behind." Byron McKinstry found the going "so steep in places that it does seem that the wagon must fall over on the cattle—much steeper than the roof of a house." The highest ridges took them up above 8,000 feet—several hundred feet higher than South Pass, and higher even than the lowest passes over the Sierra Nevada ahead. With relief, they skidded down the final slopes and walked out into the grassy Bear River Valley. Here travel would be much better. Rather than heading west across the north-south ridges and valleys, the trail now turned north, following the Bear River along its valley toward Soda Springs and the Snake River beyond. South Pass lay about 130 trail miles behind via the Sublette Cutoff, or 180 miles behind via Fort Bridger. California's gold still lay some 850 miles ahead.

THE BEAR AND
THE SNAKE

"This is the most beautiful vally that I ever beheld," James Pritchard declared as he rolled into the Bear River Valley on June 25, 1849. "I am sure the Grass cannot be surpassed by any other spot on earth. And if I were to settle anywhere beyond the pales of civilization, this would be the spot. What can be more enviting than a bold flowing river, a beautiful vally — and lofty Mountain scenary?"

The Bear River Valley — in the heart of the Overthrust Belt along the present Wyoming-Idaho border — was about as close to heaven as you could get on the trail west. "The variety of scenery and natural curiosities," Israel Hale explained, "serve in a great measure to blunt the fatigue and hardships of our trip." The valley overflowed with luxuriant grass. Dozens of clear-water streams, laden with trout, tumbled from the surrounding mountains. In emigrant days, the valley abounded "in wildfowl, ducks, geese, plovers" that William Swain and his companions enjoyed hunting "while our teams were indemnifying themselves for their past privations." Larger game also turned up, including elk, deer, wild goat, and even grizzly bear. Berry bushes by the river sagged with tasty fruit. Not since the Platte River had the going been this good — and it would not be so good again.

Today, instead of deer and grizzly bear nosing for berries along the river, you will find pumping wells, irrigation canals, and diversion dams — the infrastructure of wheat irrigation. Beyond the wheat fields, though, you still can see thousands of acres of the same tall, thick grass that the emigrants so admired. It sloshes up against the mountain slopes

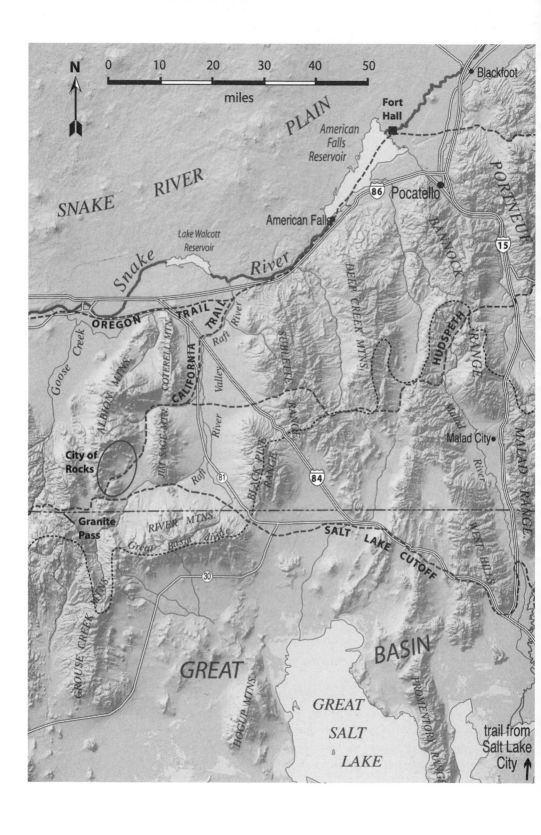

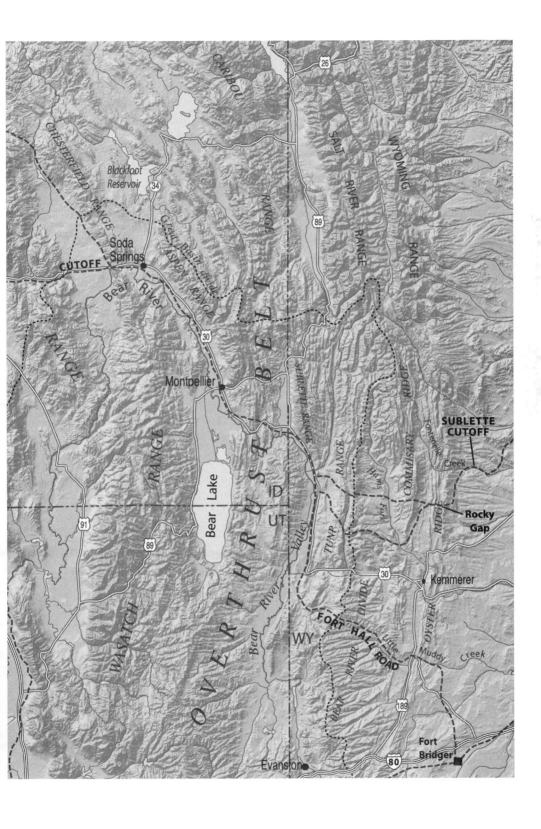

in waves of gold, rising waist high and higher, with bulging seed heads that rustle in the summer wind. From the edges of the grass ocean, the flanks of the mountains rise in steep, mottled slopes of rock and sagebrush. Pines appear high up, where the heights bring less heat and more moisture. On the ridgetops at 8,000 feet and higher, patches of snow remain late into the summer and come again early in the fall. Bent and folded rock layers—the legacy of the Sevier Orogeny—poke out from the mountainsides. Some of the layers are tilted up to vertical, their eroded edges standing like books on a shelf. Some are tilted *beyond* vertical; quadrillions of tons of rock flipped over on its back as it was shoved from the west.

9.2 Deformed rock strata in the Overthrust Belt. Sideways compression during the Sevier Orogeny took these once flat-lying layers, shoved them 30 miles east, and folded them up like rumpled towels. (In the Sublette Range a few miles east of Cokeville, Wyoming.)

9.1 (*previous pages*) After crossing the north-south ridges of the eastern Overthrust Belt, the emigrants turned north along the Bear River Valley toward Soda Springs. Beyond Soda Springs, they had to decide whether to head northwest toward Fort Hall on the Snake River or due west along the Hudspeth Cutoff. It was in this region that they got their first look at the Basin and Range Province, through which they would travel for some six weeks and 600 miles ahead. Those who took the Fort Hall Road rolled west across the vast lava beds of the Snake River Plain. Below those beds sprawl a dozen calderas—gigantic craters ripped open by some of the most cataclysmic volcanic eruptions in the known history of the Earth.

When they crossed into the Bear River Valley, the emigrants entered the drainage system of the Great Basin. Centered on Nevada and nearly twice as large, the Great Basin encompasses most of Nevada, much of Utah, and adjacent portions of Idaho, Wyoming, Oregon, and California. Internal drainage defines the Great Basin—rivers there have no outlet to the sea. They end instead at saline lakes (such as Utah's Great Salt Lake or Nevada's Pyramid Lake and Walker Lake) or pool up and evaporate at muddy, salt-encrusted sumps called sinks. The Bear River is one of the largest Great Basin rivers. It begins in the western Uinta Mountains, flows north down the Bear River Valley, and then arcs around the north end of the Wasatch Range to head south toward Great Salt Lake. All told, California-bound emigrants would log more than 600 trail miles in the Great Basin—miles that would include the most brutal stretches of their journey.

But that misery lay in the future. For now, trail life was good. The road was smooth and level, and tasty game fell regularly before crackling guns. The cool mountain air and green vistas scrubbed away the drear memories of the Green River Basin. Along 80 miles of the Bear River, there was only one major obstacle—an impassable stretch of bottomland where the river cuts through a ridge called the Sheep Creek Hills. Here the emigrants had to peel away from the river and detour several miles up and over the hills—with a fiendishly bad descent down to the river at the end (shown in fig. 9.3). "All the wheels of the wagons were tied fast, and [yet still] it slid along the ground," Margaret Frink wrote about this descent. Some emigrants kept their wagons upright "only by having fastened to the top of our loads ropes to which men clung."

Several days' travel brought them to Soda Springs—one of the scenic gems of the overland trail. Here geysers and carbonated springs bubble up for miles around. It was a singularly romantic setting, with effervescent springs misting the air to catch rainbows from the angled sun. Wandering among the springs, Margaret Frink watched water "boil up from the ground in many places, forming mounds of earth with a little cup or hollow on top. Some of the mounds are several feet in height, the water bubbling over the top on all sides." James Pritchard marveled at waters bursting up, "clear and sparkling and in many cases thrown several feet in the air. The water is constantly boiling up with a kind of hissing noise."

Some of the springs had a sparkling seltzer flavor, inspiring Franklin Starr to try some beverage experimentation. "The water requires sugar and acid and then it resembles Soda very much. We also added Ginger and drank a good deal of it." J. Goldsborough Bruff thought the water "only needed lemon syrup, to render it perfect soda water." Byron McKinstry experienced "the same peculiar tingle through the nose as one

9.3 The singular exception to easy travel along the Bear River Valley was here, where the Bear River goes through the Sheep Creek Hills. The view looks east out of Idaho toward the Sublette Range in Wyoming. The river flows west, toward you. Dynamited road-cuts today allow the highway to pass along the river. But in emigrant days it was impossible to run wagons along the bottomlands because the river meanders from wall to wall in this wide canyon. The emigrant trail therefore had to go up and over the Sheep Creek Hills, with a horrendous 1,000-foot drop back down to the river at the end (*dashed line*).

experiences when drinking" soda water, adding, "I liked it much and drank of it again and again." Some of the springs looked and tasted like beer. "They foam and blubber and are as busy as a brew-house," John Edwin Banks declared. According to Rufus Sage, drinking from these beery springs "will prove delicious and somewhat stimulating, but, if repeated too freely, is said to produce a kind of giddiness like intoxication."

The most dramatic spring was Steamboat Spring, named for its chugging noises and puffing, geyser-like sprays of water. It spurted from a volcano-like mound of travertine (a form of limestone) about five feet across and three feet high on the right bank of the Bear River, "boiling

forth," Niles Searls wrote, "with a noise resembling that made by water thrown up by the wheel of a boat." The chuffing of paddlewheel steamers on midwestern rivers was a familiar sound to many emigrants. You can't hear Steamboat Spring anymore, though. It lies drowned below 40 feet of water in Alexander Reservoir. You can see its bubbles rippling the reservoir surface from the eighth hole of the Oregon Trail Country Club Golf Course outside of today's town of Soda Springs, Idaho. There are still dozens of springs at Soda Springs, but many have vanished under the reservoir or lie sealed off under roads and housing developments.

The fizzing waters of Soda Springs, like the innumerable other springs of southeastern Idaho, exist because the Earth's crust here is fractured like shattered glass. Countless faults allow surface water to seep far underground, where it meets hot rock. That rock is mostly Paleozoic marine limestone—the petrified remains of calcareous marine algae and shelled ocean animals like corals, clams, snails, and brachiopods. The combination of faults, water, heat, and limestone makes the Soda Springs.

9.4 Steamboat Spring, sketched by J. Goldsborough Bruff on August 17, 1849. The spring, he wrote, was "named for the resemblance of the sound it gives, to that of a steamboat's paddles, under water. It is a circular tumuli of about 5 ft. diam. and about 3 ft. high; bubbling & jetting clear sparkling water, as the hissing gases escape." The cedar tree against which Bruff leaned his gun was "carved and penciled all over, as high as can be reached, with names, &c," of passing emigrants. The spring and this part of the Bear River today lie beneath a reservoir. Reproduced by permission of the Huntington Library, San Marino, California.

Under the pressures that occur thousands of feet down, underground water readily absorbs carbon dioxide to become quite acidic. The acidic waters dissolve the limestone and carry it off in solution. Warmed by the surrounding hot rock, the water—saturated with carbon dioxide and dissolved limestone—expands and rises toward the surface. You need only open a bottle of sparkling water or champagne to see how carbon dioxide, which stays dissolved in water under pressure, will bubble out when the pressure is let off. The waters of Soda Springs bubble and fizz as they surface for the same reason—depressurization that lets carbon dioxide gas escape from solution.[1] As the gas bubbles away, the water's acidity drops, allowing the dissolved limestone to precipitate as travertine. The travertine accretes around the springs as brown and gray sheets, about as hard and crunchy as peanut brittle. It builds up, layer by layer, into stepped terraces, bulbous mounds, or volcano-like cones sometimes more than 10 feet tall. Travertine is on spectacular display in downtown Soda Springs today—by the geyser. Every hour on the hour, the Soda Springs Geyser blasts soda water 100 feet into the air. The geyser never disappoints; it is fitted with a valve and automatic timer, and is thus far more faithful than Old Faithful. It's a true geyser, though—the pressure is natural, although the timing is regulated. The erupting water degasses carbon dioxide and loses acidity, precipitating new travertine, layer by layer, around the geyser's base.

A FEW MILES west of Soda Springs, the Bear River hairpins around the north end of the Wasatch Range to head south toward Great Salt Lake. Here the emigrants left the river and within a mile rolled up to a trail fork—and a decision. To the right lay the original trail to Oregon, known as the Fort Hall Road. This route took you by way of a trading post called Fort Hall on the Snake River. The miseries of this route were legendary, involving rough, hot travel across the black lava beds of the Snake River Plain. Nearly all Oregon-bound emigrants went this way. But starting in 1849, anyone headed for California had the option of taking the left fork, called the Hudspeth Cutoff. The cutoff held clear appeal for anyone bound for California, because it eliminated the long arc of the old trail up to the Snake River. From Soda Springs, the cutoff headed due west to the Raft River Valley 125 trail miles away, where it rejoined the older California Trail coming south from the Snake. "A large number of trains had preceded us on it [the cutoff], and from reliable information . . . it was a good road, and much shorter and better than by Fort Hall," J. Goldsborough Bruff noted on August 17, 1849. Like others, Bruff would soon learn that this "reliable information" about the Hudspeth Cutoff was dead wrong.

1. Nineteenth-century engineers tunneling under riverbeds learned to pump compressed air into the excavation tunnels to balance pressure from the surroundings. There is a story, possibly apocryphal, about one of the early tunnels dug under London's Thames River. As the work neared completion, the directors hosted a celebratory luncheon in the pressurized tunnel. All were peeved when the champagne poured flat, but they drank it anyway. Later, bellies full, they left the pressurized tunnel to take the air. It turned out the champagne wasn't flat after all. The CO_2 bubbled forth inside of dozens of depressurized bellies, producing a legendary onslaught of belching and gas pains.

The Hudspeth Cutoff proved to be outrageously rough. It crossed five major Basin and Range ridges over its 125-mile course. In spite of its apparent directness compared with the Fort Hall Road, the innumerable windings and up-and-down travel cut off only about 20 miles, and the rougher travel more or less negated the small savings in distance. The route also had several long waterless stretches. George McCowen, who took the Fort Hall Road, compared notes with emigrants who had taken the cutoff and concluded, "They had lost about as much time as we and reported the worst road they ever saw. Very long, steep, rocky mountains to go up and down continually. . . . All say they would take the Fort Hall Road next time." The only advantage of the Hudspeth Cutoff lay in the higher country that it crossed. The cutoff had more firewood and somewhat cooler weather than the lower, hotter Fort Hall Road.

Misconceptions about the Hudspeth Cutoff illustrate how hard it was in the mid-nineteenth-century West to get reliable geographic information. A rumor repeated in several emigrant accounts claimed that it was only 100 miles by the Hudspeth Cutoff to the Humboldt River—when in fact it was about 260. Such misinformation could be downright dangerous. Some emigrants budgeted their remaining food supplies based on overly optimistic estimates of time and distance to California, and thus started to run out of food long before journey's end. We'll visit these sad souls in the chapters ahead.

Emigrants who opted for the Fort Hall Road over the Hudspeth Cutoff had different troubles. Harsher and hotter country met them as they rolled northwest toward the Snake River, through an austere landscape of black lava beds and volcanic cones. Several days of pulling through this unwelcoming terrain brought them to Fort Hall on the Snake River, typically by mid-July to late August. (As faster parties pulled ahead of slower ones on the westbound trails, groups became more spread out; hence the wide range of arrival times at Fort Hall.)

Fort Hall was a trading post—a center of commerce for traders, Indians, and emigrants along the Snake River. Trader Nathaniel Wyeth built the fort in 1834 to trade with local Indians. The Hudson's Bay Company promptly set about driving him out of business by overbidding him in fur prices paid to the Indians and underbidding him in the prices of goods sold to the Indians. By 1837 Wyeth had no choice but to sell to the only buyer around—the Hudson's Bay Company. The company's timing was perfect. With the opening of the Oregon and California trails in the early 1840s, profits from emigrant pockets piled on top of those earned from the fur trade.

From 1841 through 1851, the Hudson's Bay Company put Fort Hall under the charge of Captain Richard Grant, "a Scotchman, from Canada,

a fine-looking portly old man, and quite courteous, for an old moun-
taineer," J. Goldsborough Bruff remembered. Margaret Frink and her
husband "were hospitably received by Captain Grant, who treated us in
a very gentlemanly manner, and formally introduced us to his wife, an
Indian woman, of middle age, quite good-looking, and dressed in true
American style. Before we left, he kindly presented us with a supply of
fresh lettuce and onions, expressing regret that because of the lateness of
the season, he had no other varieties to offer us. We thankfully accepted
them as a very unusual luxury." Supplies at Fort Hall—brought in from
Astoria on the Oregon coast—were few and dearly priced, but for the
emigrants it was buy at Fort Hall or buy nothing at all.[2]

"WE HAVE NOW reached the most northerly point of our wearisome journey," Margaret Frink sighed as she prepared to leave Fort Hall. She knew,
as did everyone else, that "the worst part of the road is yet to be passed
over."

Some of this "worst" came up as soon as they pulled out of Fort Hall
and headed west, downstream along the Snake River. The river lay to
their right—down in the depths of a sheer-walled canyon cut several
hundred feet into black beds of basalt. Beyond the river to the north, the
lava beds stretched 50 miles across the Snake River Plain. To the south,
beyond yet more jagged lava beds, rose the ragged ridges of the Basin
and Range Province—the next major gauntlet on the trail to California.
The lava plains formed a surface "so broken and split with deep chasms
that it can hardly be crossed by a man on foot." The road was appalling. Wagons bashed and creaked over lava ledges and boulders, and the
sagebrush formed an almost-impenetrable, chest-high barrier where not
beaten down by previous trains. The black basalt sucked up the summer
sun and turned the land into a roasting pit. "Heat again became oppres-
sive, the dust stifling, and the thirst at times almost maddening," Ezra
Meeker wrote of a typical day along the Snake. The mercury reached
"one hundred and twenty degrees inside our wagon," Margaret Frink
noted on July 14, 1850, although she added, "The dryness of the air, and
the high altitude, made the heat more endurable than it would have been
in a moist climate, at a low elevation."

Dry air meant dust. By July summer aridity had settled firmly over
the Snake River Plain. Soils had cracked and shriveled, and grasses had

2. The original site of Fort Hall lies on the south bank of the Snake River about 15
miles northwest of present Pocatello, Idaho, just upstream from where the river enters
the American Falls Reservoir. The original fort no longer exists. The "Fort Hall" that trail
buffs and tourists visit today is a replica built near downtown Pocatello, miles from the
site of the original fort.

9.5 The Snake River winding through its deeply entrenched canyon. Although the Snake was the biggest river the emigrants had seen since the Missouri, in many places its high, sheer walls made access to water difficult. The view looks downstream a few miles west of American Falls, Idaho.

turned raspy and dry. Grazing and trampling laid the ground bare, and dust rose at the slightest disturbance. "It will fly so that you can hardly see the horns of your tongue yoke [the oxen closest to the wagon]," Oregon-bound Elizabeth Dixon Smith wrote in August 1847. "It often seems," she added, "that the cattle must die for the want of breath, and then in our waggons such a specticle beds, clothes, vituals [victuals?] and children all completely covered."

At least once a day along the Snake, emigrants had to find a way to drive their cattle to water—which meant finding a way down the precipitous walls of the Snake River Canyon. If they ended the day at a waterless campsite, they had to scramble down to the river and haul water up in buckets. One creature had no difficulty making the trip between the river and the bluff-top campsites. Israel Hale met with mosquitoes "more numerous than I have ever seen them, and hungry as wolves," while Margaret Frink slapped at clouds of mosquitoes "as thick as flakes in a snowstorm." James Pritchard remembered Independence Day—July 4, 1849—along the Snake with this entry:

> Our Fourth of July was spent traveling in the dust and fighting off Mus-
> ketoes. Their attacks were more fearce and determined, and more nu-

merous, along this river than any of the kind I ever witnessed. . . . The
Musketoes were so bad this evening that they nearly ran Our mules
crazy. They would break & run & lay down & roll, jirk up in picket pines
or break Lorrietts. We cut sagebrush & grees wood & built large fires all
round the camp—raised such a smoke that it would suffocate the men
& it did no good.

Tempers broke and ran before such tribulations. The accretion of
hardships inevitably wore down spirits and civility. Israel Lord reached
deep into his bag of adjectives to describe the transformation. Men had
become, "by turns, or all together, cross, peevish, sullen, boisterous,
giddy, profane, dirty, vulgar, ragged, mustachioed, bewhiskered, idle,
petulant, quarrelsome, unfaithful, disobedient, refractory, careless, con-
trary, stubborn, hungry and without the fear of God and hardly of man
before their eyes."

∗

The Snake River heads in the high country of the Yellowstone Plateau, a
stone's throw west of the Continental Divide. It pauses in Jackson Lake
at the foot of the Teton Range, and then tumbles south through Jackson
Hole before turning west to punch through the Snake Range. From there
it emerges onto the volcanic plain that bears its name, cruising west along
the south edge of the Snake River Plain to join the Columbia River 700
miles downstream.

The Snake River Plain cuts a wide scar across the face of the West, like
a wound where the mountains never grew back. It forms a flat-bottomed,
east-west-trending trough, 40 to 60 miles wide, that plows at right angles
across the regional north-south grain of the mountains. Your eye trips
over it. Why is it there? The answer lies to the east. Follow the Snake
River Plain east-northeast into Wyoming, and you climb directly into the
steaming world of Yellowstone National Park.

In the latest blink of geologic time, 640,000 years ago, a gargantuan
volcanic blast ripped a 45-mile-wide pit in the Earth's crust in present
northwestern Wyoming. Today it forms the heart of Yellowstone, and
residual magma from that blast, close underground, powers Old Faithful
and the other geothermal features of the park.

A string of similar eruptions, some even more powerful, gouged out
the vast trough of the Snake River Plain. Today the Snake River Plain is
chock-full of black volcanic basalt. But these lava beds are just a veneer
pasted over a much more violent volcanic past. Oozing from deep fissures as recently as 2,000 years ago, the basalt lava spread like honey in

a gutter, filling in a vast ditch blasted out by earlier eruptions. The result is today's landscape of flat, black lava plains. Where the Snake River cuts its canyon down through these lava beds, it exposes a wholly different history—represented by rhyolite tuff. Powdery and soft, colored in pastel shades of yellow, pink, and tan, rhyolite tuff is compacted volcanic ash—tiny jagged particles of rock and volcanic glass with larger rock chunks mixed in. The ash blasted out from a 500-mile-long string of volcanic centers that now lie mostly hidden below the lava beds of the Snake River Plain. The easternmost of these volcanic centers lies below Yellowstone.

The rhyolite tuff of the Snake River Plain speaks of volcanic hell, complete with fire. Scalding clouds of ash and red-hot rock particles surged repeatedly from the volcanic centers along the Snake River Plain, leveling forests and instantly turning every nearby animal into burnt meat. Some of the ash blew sky-high, eventually landing as far east as the Great Plains.[3] Emptied of magma, the eruption centers collapsed to form great craters, or calderas. Calderas are yawning holes, tens of miles across, formed where large volcanoes empty themselves in wrenching magmatic orgasms before collapsing into the underground spaces where the magma used to be. The calderas of the Snake River Plain drape like a string of beads across the neckline of Idaho, and just as beads join up one by one to make a necklace, so the calderas formed one by one, in sequence. The oldest caldera—16.5 million years old—lies farthest to the west, in southeastern Oregon. The calderas become younger as you track them eastward across Idaho toward Yellowstone (fig. 9.6). At the national park named for its ubiquitous yellow stone (rhyolite tuff) you find the youngest caldera in the sequence—the Yellowstone Caldera.

A trip through Yellowstone National Park will not show you an obvious caldera. That's because nearly everywhere you go in the park you are *inside* the gigantic Yellowstone Caldera. The 45-mile-wide volcanic pit is simply too big to take in from the ground. Even from the air, it doesn't jump out as a clear-cut crater (unlike, say, Oregon's Crater Lake Caldera or Hawaii's Kilauea Caldera). Events since the caldera's formation have obscured its crater-like form. Rising heat has pushed up a series of bulges called resurgent domes in the middle, and lava eruptions have filled in much of the rest of the caldera to form the broad plateau on which most of Yellowstone's bears, buffalos, and tourists presently tread.

In spite of these obscuring developments, geologic mapping has shown that the greater Yellowstone region houses not one but *three* overlapping

---

3. You may recall from chapter 4 the story of the 12-million-year-old fossil rhinoceroses buried in volcanic ash on the Great Plains. The ash that doomed those rhinos to starvation and suffocation erupted from a caldera near Pocatello, Idaho, on the Snake River Plain.

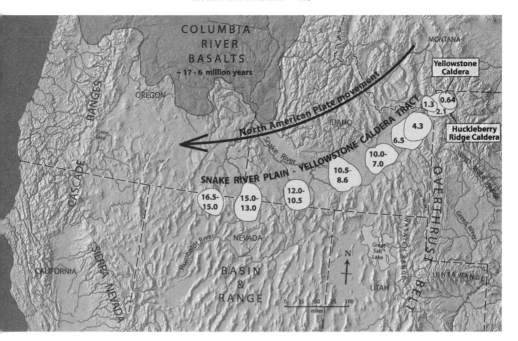

9.6 The 500-mile-long Snake River Plain–to-Yellowstone caldera tract begins in southeastern Oregon and northern Nevada with the oldest caldera and ends at the active Yellowstone Caldera in northwestern Wyoming. The numbers show the ages of caldera-forming eruptions in millions of years before present. Notice the clear east-west age progression. Most of the calderas have been sites of multiple eruptions; hence the range of ages at some sites. The age-progressive trend may reflect the west-southwest movement of the North American Plate over a stationary mantle plume (although recent research now questions this interpretation). The eruption of the massive Columbia River Basalt field, to the north, coincides with the crustal stretching that formed the Basin and Range (discussed in chapter 10), suggesting that the basalt lava erupted through fissures torn during crustal stretching. The relationship (if any) to the Snake River Plain–to-Yellowstone caldera tract is unclear.

calderas, corresponding to three titanic eruptions. The most recent one—the 640,000-year-old blast that made the present caldera—hurled 240 cubic miles of rock and magma into the sky—enough to fill Lake Erie two times. Ash from that eruption landed as far away as Iowa and Louisiana. The eruption before that, 1.3 million years ago, was a more modest 67 cubic miles—still 700 times larger than the 1980 eruption of Mount Saint Helens, whose ash clouds killed fifty-seven people and brought the Pacific Northwest to its knees for weeks. The eruption before that—the 2.1-million-year-old Huckleberry Ridge eruption—was one of the largest volcanic blasts in the known history of the Earth. It heaved an estimated 585 cubic miles of rock and magma into the stratosphere (Mount Saint Helens 6,000 times over), slinging ash west to the Pacific Ocean, east

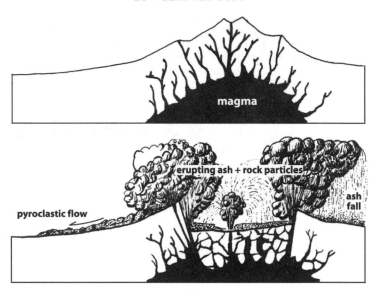

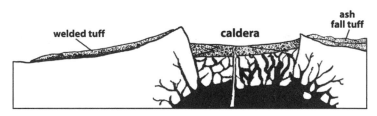

9.7 Volcanic calderas form when volcanoes erupt with such violence that the central cone collapses into the partially emptied magma chamber below the volcano, forming a crater many miles across. Erupted material blasts out in two major ways: as dense, red-hot clouds called pyroclastic flows that rush downhill like dense, scalding avalanches, or as particles that rise many miles into the air and then fall back as ash falls. Both pyroclastic flows and ash falls form a rock called tuff. The form of tuff made by pyroclastic flows is called welded tuff because the glowing hot particles weld themselves together into a harder, denser rock than ash fall tuff, which is usually powdery and soft. Reproduced by permission of the University of Utah Press.

to Minnesota, and south to Mexico. Spread out evenly, the ash from the Huckleberry Ridge event alone would have buried all of the states west of the Mississippi River two feet deep. No eruption in recorded human history even comes close. The 1883 eruption of Krakatoa in Indonesia—which decapitated a five-mile-wide island and killed more than 36,000 people—was 100 times *smaller* than Huckleberry Ridge.

And that's just the last 2 million years at Yellowstone. When you add the Snake River Plain calderas to the mix, you are looking at a 16.5-million-year history of comparably large eruptions—one that began in southeastern Oregon and advanced east 500 miles, blast by blast, across southern Idaho to Yellowstone. On the face of it, it looks as if a deep source of magma has migrated east 500 miles underneath North America, detonating calderas one after another in sequence for the past 16.5 million years. But this idea probably turns the process on its head. It turns out that 500 miles divided by 16.5 million years equals nearly two inches per year—close to the rate at which seafloor spreading pushes the North American Plate west from the Mid-Atlantic Ridge. In other words (according to current theory), the Snake River Plain–to–Yellowstone caldera tract formed not because a magma source under North America migrated *east*, but because North America migrated *west* over a magma source that stayed more or less stationary. Imagine lying on your back underneath a trailer home holding a sawed-off shotgun, pointing up. As the trailer rolls away, you squeeze off a series of shots. Someone inside the trailer sees the floorboards explode in a line of holes from one end to the other. The trailer is the North American Plate, the floorboard holes are the line of Snake River Plain–to–Yellowstone calderas, and the stationary shotgun represents a mantle plume—a column of incandescent rock rising up through the mantle. As the North American Plate slides along, the mantle plume beneath it periodically uncorks a caldera-forming eruption.

According to this widely held theory, the mantle plume that spawned the Snake River Plain–to–Yellowstone caldera tract now lurks below Yellowstone—reloading. The plume's heat powers the geysers, steam vents, hot pools, and bubbling mud pots of the park. Heat pours out of the crust at Yellowstone at rates 30 times higher than the continental average. The heat expands the crust, elevating it into a broad plateau 7,500 feet above sea level—with the caldera smack in the middle. Masses of magma, shifting and gathering just a few miles below the caldera, make Yellowstone one of the most seismically active places on Earth. Anywhere from 1,000 to 3,000 earthquakes rattle the park each year. Most are so small that only sensitive seismometers pick them up, but some are big enough to shake soda cans on picnic tables and a few are large enough to trigger landslides. The largest quake in the park's history, in 1959, tore a 22-foot-high scarp (a cliff formed where a fault shifts) through the park and unleashed a landslide that killed 26 people. The landslide dammed the Madison River to make Earthquake Lake—the youngest natural lake in the United States. From 1923 to 1984, the center of the caldera rose three feet. Over the next ten years, it sagged back eight inches . . . and then began rising again. The land heaves like a dragon's chest, mirroring the restless magmatic world below. Yellowstone has had 640,000 years to gear up since its latest titanic blast—a number converging on the average time gap between previous eruptions. It will explode again one day (probably east of the present caldera), with similarly cataclysmic results. That would be a good time to be far, far away.

MOST OF THE Earth's volcanoes form at the edges of tectonic plates, bursting up through continental rift valleys and mid-ocean ridges, or erupting in volcanic arcs above subduction zones. But the Snake River Plain-to-Yellowstone caldera tract lies nowhere near the edge of the North American Plate. Yellowstone is one of the world's classic hot spots—point sources of intense and long-lived volcanic activity that usually lie far from any plate boundary. As a geologic hot spot, Yellowstone joins a club that includes Hawaii, Iceland, the Galapagos, the Azores, Tahiti, Easter Island, and more than a dozen others.[4]

The key thing about hot spots is that they anchor lines of volcanoes or calderas that get progressively older with distance from the hot spot. In 1971 Princeton University's W. Jason Morgan proposed that hot spots form over mantle plumes, which he envisioned as narrow columns of super-hot rock rising from deep in the mantle, perhaps from the core-mantle boundary 1,800 miles down. These plumes, according to Morgan, rise up until they bump into the bottoms of plates, where they spread out like thunderheads. The release of pressure on the rising rock lets some of it melt, squirting up to form volcanoes or calderas. Plates move but mantle plumes stay put, Morgan suggested. Consequently, a plate moving over a mantle plume will be adorned with a trail of progressively older volcanic cones or calderas downstream of the active hot spot.

In Morgan's theory, the 3,600-mile-long Hawaiian-Emperor chain of volcanic islands and seamounts (undersea volcanoes) is a classic example of a mantle plume at work. The seabed around the big island of Hawaii bulges up hundreds of feet into a broad swell about 1,000 miles across—an effect, presumably, of the immense heat of the plume below.[5]

4. While most hot spots occur far from plate boundaries, there are exceptions. The Iceland, Azores, and Tristan hot spots all sit on the Mid-Atlantic Ridge, where the seafloor spreads apart. The coincidence of some hot spots with spreading ridges suggests that the source of heat that powers hot spots may also play a role in breaking up continents and triggering seafloor spreading. Like the perforations in sheets of postage stamps, hot spots may weaken continental plates and so control where they tear apart to form new ocean basins.

5. Hawaii sits at the center of this 1,000-mile-wide seafloor bulge, but the massive weight of the volcano pushes the center of the bulge down, forming a gentle "moat" of deeper seafloor all around the island, like the depression formed around you when you sit on a waterbed. Hawaii starts 18,500 feet below sea level and tops out 13,800 feet above sea level at Mauna Kea, making it the world's tallest mountain if measured from the ocean floor.

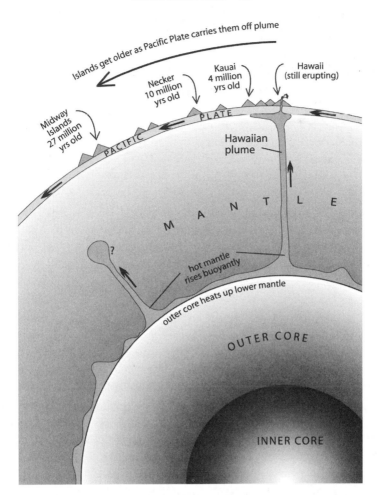

9.8 The Hawaiian-Emperor chain of volcanic islands and seamounts (undersea volcanoes) is the poster child of traditional mantle plume theory. The theory holds that volcanoes erupt through the Pacific Plate above the plume, after which the conveyor belt-like movement of the plate carries them northwest off the plume. The purported plume presently sits under the big island of Hawaii. The theory suggests that mantle plumes arise from the core mantle boundary where intense heat from the outer core forms buoyant blobs of super-hot rock. Such deeply rooted mantle plumes have thus far evaded detection by seismic tomography—the "X-ray" method we use to peer into the mantle. (The Pacific Ocean is too thin to appear on the scale of this diagram, and the size of the volcanic islands is hugely exaggerated.)

Over the past half-million years, magma from the plume has built the big island, layer by lava layer, up from the seafloor. Lava presently oozes almost without letup today from Kilauea volcano on Hawaii's southeast flank. But the Pacific Plate, moving northwest several inches each year, is carrying Hawaii off the plume. Consequently, even though Kilauea still bubbles red-hot, lava also erupts from a new volcano, called Loihi, underwater 30 miles southeast of Kilauea. Growing bigger with each eruption, Loihi is making a bid to become the next island in the Hawaiian chain. The older island offspring of the Hawaiian plume march off to the northwest: Maui (800,000 years old), Molokai (1.8 million years old), Oahu (2.5 million), and Kauai (4.5 million).[6] The aging continues northwest to Necker Island (10.3 million) and the Midway Islands (27 million years old). Beyond Midway, none of the volcanoes in the chain breaks the ocean surface. The line turns north at a dogleg bend (43 million years old) to become the Emperor Seamounts, including Koko (49 million), Suiko (61 million), Detroit (75 million), and Meiji Seamount—81 million years old and 3,600 miles from Hawaii. Beyond Meiji, the Pacific Plate ends, bending down into the mantle at the Aleutian and Kuril trenches.

Morgan's mantle plume theory for hot spots and age-progressive lines of volcanoes is magnificently simple and elegant. It is so widely accepted that it has become virtual gospel. You would be hard-pressed to find a geology book from the last four decades that does not explain the Hawaiian-Emperor volcanic chain, the Snake River Plain-to-Yellowstone caldera tract, or any other age-progressive line of volcanoes this way. Textbooks confidently portray mantle plumes rising through the mantle like 1,800-mile-long cobras of incandescent rock, with their tails anchored on the outer core (the presumed source of heat that spawns plumes) while their heads feed volcanic fire to the plates above. There's just one problem— we don't know if it's true. We *see* lines of volcanoes or calderas that get older with distance from active hot spots; we *infer* that they result from plates sliding over deeply rooted, stationary mantle plumes. Now some renegade geologists are opening fire on this cherished theory. And Hawaii and Yellowstone lie squarely in the crosshairs of the debate.

For decades geologists have had the Hawaiian plume firmly anchored in the mantle, fixed and unmoving, spouting molten rock while the Pacific Plate slid by overhead. The 43-million-year-old dogleg bend in the

6. The ancient Hawaiians knew that their islands got older going northwest. They could tell from the relative freshness of the lava and the degree of erosion and soil development—as can any observant person today, for the islands clearly look softer, more vegetated, and more eroded as you island-hop northwest. The Hawaiians had an explanation. Pele, the volcano goddess, originally lived on Kauai. But her bullying older sister Namakaokahai, the goddess of the sea, chased her first to Oahu, then Maui, then Molokai, and then onto the big island of Hawaii. Being a volcano goddess, Pele paved each island she visited with lava. Today she lives in Halemaumau Crater on Kilauea volcano in southeastern Hawaii. Pele has written part of geology's lexicon; teardrops of volcanic glass that congeal during spattering eruptions make "Pele's Tears," and thin strands of volcanic glass make "Pele's Hair."

Hawaiian-Emperor chain, the story went, happened when the Pacific Plate shifted direction. Why did that happen? No one knew. How did the Pacific Plate change direction without affecting any of the plates next door? No one knew that either. But there was that bend, by god, and other bends like it in other lines of Pacific volcanoes. The Pacific Plate *must* have shifted direction to make those bends, since mantle plumes are stationary—right?

Maybe not. Recent sampling of the volcanic basalts on the Emperor Seamounts suggests that the Hawaiian plume may have drifted south *by hundreds of miles* while it was forming the seamounts.[7] So much for stationary mantle plumes. Perhaps plumes wander around like chimney smoke in a breeze. That breeze might be the mantle wind—a term sometimes used to describe the slow, roiling currents of convecting hot rock in the Earth's upper mantle. If a strong mantle wind pushed the Hawaiian plume south during the formation of the Emperor Seamounts, there may be no need to have the Pacific Plate make that odd directional shift 43 million years ago.

What about the mantle plume below Yellowstone? No one questions the age-progressive nature of the Snake River Plain-to-Yellowstone caldera tract—oldest to the west, youngest to the east. Taken by itself, the tract seems to fit perfectly with the idea of the North American Plate sliding west over a stationary mantle plume. But not so fast. Recent work has revealed another age-progressive volcanic tract nearby. The Newberry tract begins in southeastern Oregon (where the Snake River Plain-to-Yellowstone tract also begins)—but it gets younger to the *northwest*, completely at odds with the Snake River Plain-to-Yellowstone tract. Since the North American Plate can't be going in two directions at once, we can't explain both age-progressive volcanic tracts with a single stationary mantle plume. So what exactly *is* under Yellowstone? Is there even a mantle plume there at all?

Seismic waves are the only tool we have to see into the mantle where plumes supposedly lurk. Just as computerized tomography (CT scan-

7. The evidence for southward movement of the Hawaiian plume comes from measurements of the latitude of the Emperor Seamounts when they erupted. If the Hawaiian plume has stayed in one place, then the seamounts should have formed at the latitude where the plume is today, under Hawaii 19 degrees latitude north of the equator. You can measure the latitude where lava formed because the vertical angle of the Earth's magnetic field changes going from the equator to the poles, and the magnetic minerals in lava rock record this angle when the rock solidifies. Magnetic measurements in basalts collected from the Emperor Seamounts seem to show that the seamounts originally formed as far north as 37 degrees north latitude—a result impossible to reconcile with a stationary mantle plume. See the Notes for references.

ning) images a person's insides using crisscrossing waves of X-rays, so a method called seismic tomography takes grainy pictures of the Earth's mantle using multiple seismic waves. The Earth's grinding tectonic plates fire off thousands of earthquakes every year. The waves from these quakes form a crisscrossing web as they pass through the mantle. Seismic waves slow down markedly through hot, mushy rock—a key feature for detecting mantle plumes. To find plumes, geologists look for these slow-down regions. The problem is that the mantle is a colossally large place— 1,800 miles thick and making up 80 percent of the Earth by volume. Seismic tomography has so far mapped out only a general picture of mantle temperatures. We can see haystacks of hot and cool rock many hundreds of miles across, but mantle plumes are more like needles. If we had millions of seismometers scattered evenly over the planet, gathering the daily traffic of seismic waves over many years, then mantle plumes—if they exist—should pop up on computer screens like erupting fire hydrants. But such a global density of seismic sampling is beyond our reach at present.

One exception is Yellowstone. The greater Yellowstone region boasts one of the densest seismograph arrays on Earth. It shows us an area of exceptionally hot rock, 60 miles wide, poking down about 300 miles below Yellowstone. And that's all. Below that, the region of super-hot rock tapers away to nothing. Seismic imaging below Hawaii and Iceland (while not yet as refined as Yellowstone) likewise seems to show nothing notable below about 300 miles.[8] Where are the mighty mantle plumes—those 1,800-mile-long cobras of incandescent rock that rise so confidently from the pages of every geology textbook? Not, apparently, inside planet Earth.

Mantle plume theory—at least in the sense that Morgan originally proposed it—is in trouble today. We thought that plumes were stationary. Now evidence suggests that plumes—if they exist—don't stay put. They sway and drift in the mantle wind. We thought that plumes began deep, at the core-mantle boundary 1,800 miles down. But we can't find anything plumelike more than 300 miles below either Hawaii or Yellowstone.

8. Some geophysicists, teasing global seismic data in new ways, claim to have detected plumes going all the way through the mantle. Others doubt the resolution and assumptions of these methods, claiming, in effect, that haystack-level sampling can't reveal needles. Soon an ambitious project called PLUME (Plume-Lithosphere Undersea Melt Experiment) may better resolve the depth and size of the Hawaiian Plume. The project is gathering seismic data from an array of portable seismometers scattered for thousands of miles across the seafloor around Hawaii. This density of instrumentation should achieve an unprecedented level of imaging detail for the mantle beneath Hawaii.

When a theory fails to yield evidence of its most basic predictions, you have to wonder whether it's time to file for divorce. It may be time to pull mantle plume theory out of our collection of ideas about how the Earth works. But what about age-progressive lines of volcanoes or calderas trailing off from active hot spots? Don't we need mantle plumes to explain those? Maybe not. An alternative theory holds that age-progressive volcanic lines could form along slowly propagating *cracks* in tectonic plates. Plates are big and experience monumental stresses; we should expect them to crack. Like a fracture creeping across a windshield, a propagating crack might let volcanoes pop up one after another at its leading edge. Such cracks might not even show up at the Earth's surface; they might simply be deep, migrating zones of weakness along which magma could tunnel upward and burst out. The reaction to *crack theory* as a viable alternative to *plume theory* has been mixed, with predictable renaming (cracked theory, crackpot theory). But crack theory has fully joined the debate over what might have made the Hawaiian-Emperor chain and the Snake River Plain–to–Yellowstone caldera tract. The arguing and discussion, the questioning of cherished assumptions, the hors d'oeuvres of opinion flavored with dashes of vitriol—all represent science at its best. Science has the habit of dismantling its own house now and then, and better theories eventually arise from the wreckage of broken paradigms. It's all part of the fun and challenge of prying knowledge from a reticent Earth.

A BREAKING UP OF THE WORLD

For California-bound emigrants, the dismal trek along the volcanic plains of the Snake River ended where the Raft River joins the Snake from the south. Here the trail forked. Everyone headed for California turned south up the Raft River Valley, leaving those bound for Oregon to continue west along the Snake. "'The Oregon Trail' strikes off to the right & leaves us alone in our glory, with no other goal before us but Death or the Diggins," Wakeman Bryarly noted at this parting point. The Humboldt River—the next major goal on the road to California—lay 160 miles ahead. The Humboldt would be the emigrants' lifeline across the arid Great Basin, taking them within striking distance of the Sierra Nevada. "If no bad luck happens," Israel Hale wrote at Raft River, "we will, in thirty-five or forty days, reach the land that is said to abound in gold."

You will find no Raft River in the Raft River Valley today. Emigrants described the Raft as a pretty stream, about 10 feet wide, gravelly and swift, lined with willows and reeds. Today it is a mud-cracked trench clogged with shriveled weeds and beer cans. Dozens of irrigation wells have inhaled the groundwater that once nourished the river, converting it into thousands of acres of hay. The dead river lies in a sun-baked valley spreading 20 miles between flanking ranges: the Sublette and Black Pine ranges to the east, the Cotterell and Jim Sage ranges to the west. Range, basin, range. Stepping into the Raft River Valley, the emigrants once again entered the Basin and Range geologic province.

The immense Basin and Range Province sprawls from central Utah and westernmost Wyoming west across Nevada to California, and from

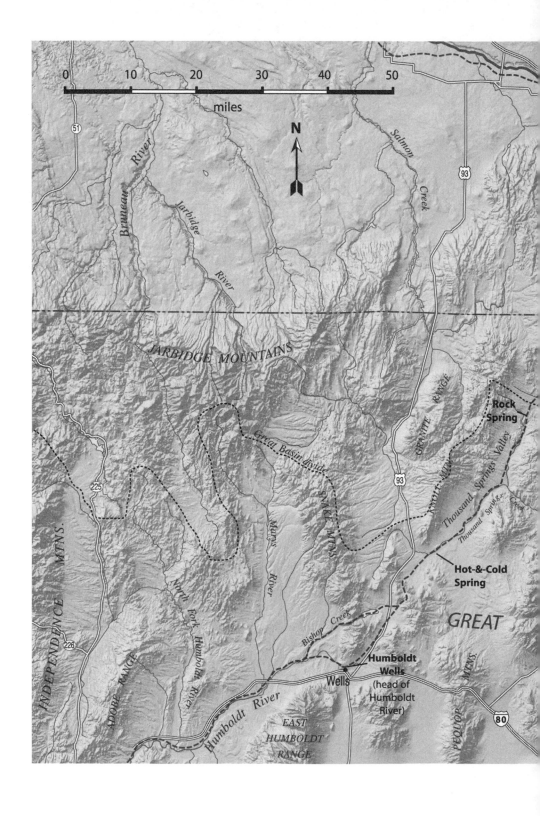

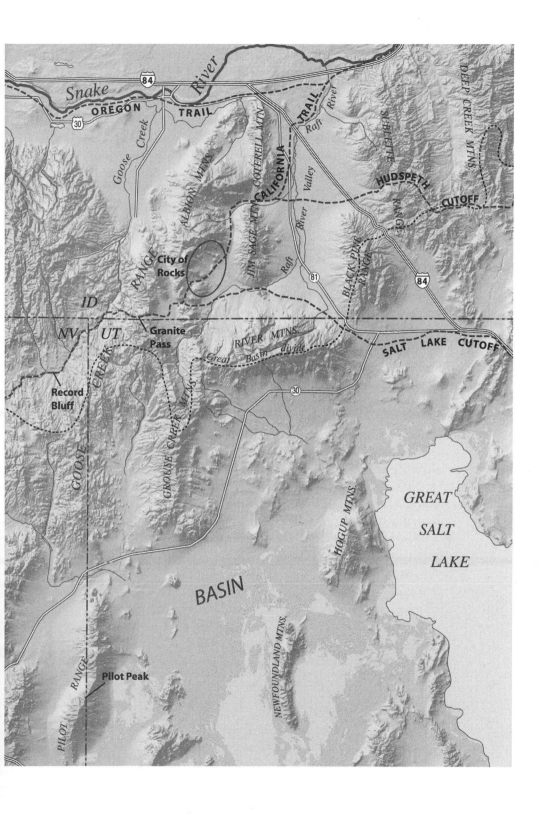

southern Oregon and southern Idaho south far into Mexico. Overland emigrants to California—whichever way they went (along the original trail south from Snake River or over the Hudspeth, Salt Lake, or Hastings cutoffs)—would travel some 600 miles through this broken land of alternating basins (valleys) and mountain ranges. You can't go anywhere in the Basin and Range and be out of sight of mountains. The province contains some 500 ranges, most of them long (40 to 100 miles), narrow (10 to 20 miles), and lined up north-south. The Basin and Range is high country. Many of the basins lie 4,000 to 7,000 feet above sea level (higher than most of the United States east of the Rockies), and some of the ranges reach over 13,000 feet.

For the emigrants, the north-south mountain barriers of the Basin and Range stood like rows of abatis against their westward advance. Frontal assaults on these fortifications were next to impossible. Instead, the twisting overland trails outflanked the mountains, snaking around the highest ridges, following the routes of meager streams. Only where necessary did the trails confront the ranges head-on, sneaking over saddlelike passes to get to the next basin. The reward for running this gauntlet of mountain ramparts was to reach the walls of the castle—the Sierra Nevada.

Within the Basin and Range, and taking up the better part of it, lies the Great Basin. The emigrant trail through the Basin and Range also crosses the heart of the Great Basin. Although the term "Great Basin" suggests a bowl-like plain surrounded by high rims, the region is no less mountainous than any other part of the Basin and Range. The Great Basin's borders are defined by divides that separate streams flowing in from those flowing away. The Sierra Nevada forms the western divide. The Wasatch Range and the high plateaus of south-central Utah form the eastern divide. To the north and south, the basin's borders are marked by low ridges that separate inward-flowing streams from those that link up with the Snake and Columbia rivers to the north or the Colorado River to the south.

For the emigrants, the Basin and Range/Great Basin would dish out the harshest tribulations of the westward journey. Here geologic circum-

10.1 (*previous pages*) Three separate California-bound trails converged south of the Snake River. The original California Trail south from the Snake joined up first with the Hudspeth Cutoff and then with the Salt Lake Cutoff. All emigrants then headed for the Humboldt River. The trail routes shown are within the Basin and Range geologic province, where east-west stretching of the crust has broken the land into north-south-trending mountain ranges and valleys (basins). As the land stretched, the central region sagged to form a vast area of inward drainage, forming the Great Basin. California-bound emigrants would travel 500 miles within the Great Basin, all the way from the divide shown here in northeastern Nevada to the summit passes of the Sierra Nevada.

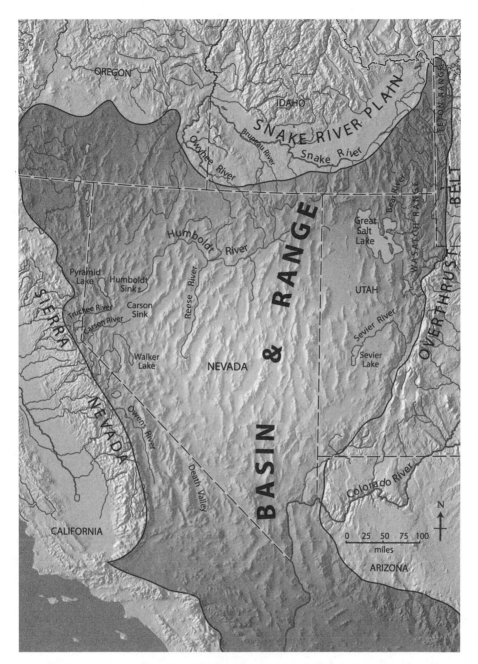

10.2 The northern portion of the Basin and Range Province. The province extends well south and east of the region shown here, reaching across Arizona into New Mexico and south far into Mexico. The province is defined by north-south-trending ranges and basins formed by east-west crustal stretching during the Basin and Range Orogeny, which began about 20 million years ago and continues today.

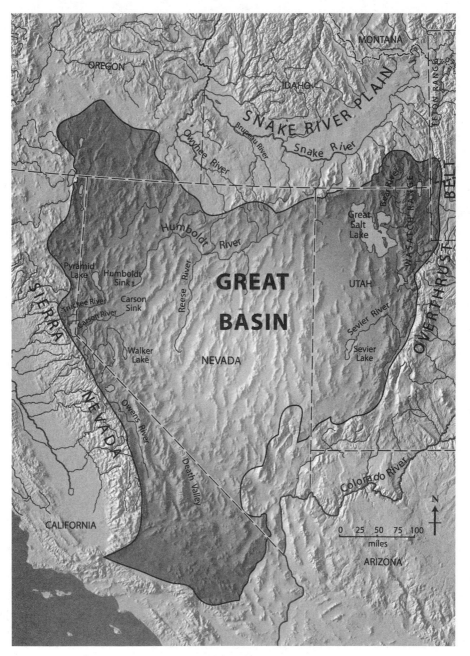

10.3 The Great Basin sits within the northern Basin and Range Province and takes up the better part of it. As the crust stretched during the Basin and Range Orogeny, it sagged in the middle to form this vast region of inward drainage. Great Basin rivers begin and end in the Great Basin instead of flowing to the sea. The 350-mile-long Humboldt is the longest Great Basin river.

stances would mate with summer's heat to spawn misery, thirst, hunger, abandonment, and death in frightful abundance. How different the trek to California would have been if the landscape of the American West had come in reverse, with the Basin and Range coming first and the easy-going Great Plains at the end. It would have been an easier journey. But geologic history, unfolding as it did, set the course of emigrant history as we know it.

THE SNAKE RIVER fell behind as the emigrants rolled south up the Raft River Valley. But cloying dust—one of the banes of the trail along the Snake—remained. The dust came from beds of rhyolite tuff—compacted volcanic ash spewed out of the Snake River Plain calderas. Boots, hooves, and wheels pulverized the soft volcanic rock so that "the least breeze or trampling of the cattle is sufficient to raise it in such clouds as to envelop a whole train," James Bennett observed, adding that the dust could be so thick as to "render it difficult to distinguish the wagons at a few yards distance. Sometimes, too, a whirlwind will come sweeping up a ravine, carrying the dust spiraling mountain high, almost smothering teams and teamsters in its passage."

Conditions improved upstream, to the south. The air cooled as the valley rose. Juniper trees appeared, providing firewood and a break from the stark and treeless plains. The Raft River yielded a new food: cray-fish—exotic fare compared to the monotonous trail staples of bread and bacon.[1] These "fresh water lobsters are boiled and eaten by many, said to be good."

Some 30 miles up the valley, a line of dust approaching from the east signaled the arrival of emigrants on the Hudspeth Cutoff. The two trails, which had split more than 100 miles back near Soda Springs, now re-joined. United, all emigrant traffic now turned southwest up the valley of Cassia Creek, a tributary of the Raft River. The road steepened. Pine- and juniper-dotted ridges rose up all around. Then, where the trail turned up a valley into the Albion Mountains, a crescendo of wild mountain scen-ery burst into view. Solemn monoliths of granite, hundreds of feet tall and eroded into fantastic shapes, reared up across a bowl-shaped valley several miles wide. The emigrants rolled to a stop, bedazzled. They had come to the City of Rocks.

---

1. Nearly all emigrants counted on bread and bacon to get them to California. The two staples kept the body going, but the soul tired of the unchanging fare. "How we do wish for some vegetables," emigrant Helen Carpenter mourned, adding, "One does like a change and about the only change we have from bread and bacon is to bacon and bread."

10.4 City of Rocks in southern Idaho. The emigrant trail passes for several miles through this surreal collection of granite monoliths. Many emigrants compared the scene to that of a silent city. For Margaret Frink, it was a "sublime, strange, and wonderful scene—one of nature's most interesting works."

California-bound emigrants had by now seen many spectacular rock formations: Courthouse Rock, Chimney Rock, Scotts Bluff, Independence Rock, Devils Gate, and others. But these were all single monoliths, spectacular yet solitary. At the City of Rocks, the granite towers and domes loomed up by the score. "You can imagine among these massive piles, church domes, spires, pyramids, &c., & in fact, with a little fancying you can see [anything] from the Capitol at Washington to a lowly thatched cottage," Wakeman Bryarly marveled. William Swain saw all the components of a silent city—"a mass of common buildings, the streets, the town pump, the taverns and their chimneys, the churches with their spires, the

monuments of the graveyard, the domes, the cornices and the columns. These all had their representations in this wild scene of nature."

CITY OF ROCKS National Reserve lies in an empty quarter of America, in backcountry near the intersecting borders of Idaho, Utah, and Nevada. The reserve is inconveniently far from everything, yet visitors pour in here in astonishing numbers. Many are rock climbers. They swarm over the monoliths, dangling high above the ground, trusting limb and life to the uncertainty of rugose granite. Some are rut nuts like me. We wander the sagebrush flats, wiping sweat from our eyes, searching for wispy traces of old wagon trails. Then there are the sensible people—the ones who head for shade among the craggy rocks and unpack sumptuous lunches, untroubled by imminent death or dehydration.

City of Rocks began more than 30 million years ago as gobs of magma welling up through the crust below present southeastern Idaho. The magma stalled about eight miles down, where it congealed into the Almo Pluton. As the crust stretched in the early stages of the Basin and Range Orogeny (more about this in the pages ahead), the rock above the pluton slid away to the east like blankets drawn off a bed. It came to rest about 50 miles away, where it now forms the Sublette and Black Pine ranges east of the Raft River Valley. As the rock overhead slid away, the Albion Range (in which the Almo Pluton resides) bobbed up like a slow-motion cork.[2] Swelling from the release of pressure, the pluton split into numerous decompression cracks, or joints—like those we saw in the Sweetwater Hills (chapter 6). These joints hold the key to the City of Rocks.

If you have ever repaired a driveway or foundation, you know how water wreaks havoc wherever it penetrates cracks. As the Almo Pluton bobbed upward, underground water seeped into the joints, disassembling the connections between atoms, dividing mineral from mineral. Where two or three joints intersected to form corners and edges, the water attacked from two or three directions, deteriorating those areas fastest. In this manner, the mass of granite—still underground but thoroughly infiltrated by water along countless joints—slowly changed from a collection of close-fitting angular blocks into a group of semi-round monoliths surrounded by rotted granite debris. Erosion then slowly worked its way down through the split-up granite mass. It scoured away the disintegrated debris along the joints, revealing the collection of rounded, fractured granite blocks that we see today at City of Rocks.

Exposed to sky and weather, the monoliths then fell prey to another trick of water—one that vastly enhanced their grotesque and winsome shapes. In the arid climate of southern Idaho, rainwater on rock surfaces rapidly evaporates. It leaves behind thin mineral coatings that actually harden rock surfaces exposed to rain. This process—called casehardening—lets some rock surfaces withstand further erosion while other surfaces continue to molder and peel away. The result is numerous overhanging caps, cavities, tunnels, and caves, all conspiring to create a scene that—for 1850 emigrant Leander Loomis—"fills the mind of man, with a wild romantic Grandeur, which raises him above his natural sphere, and leads him to aspire, to reach the angles [angel's] seat."

2. Readers with some geologic background may recognize that this makes the Albion Range a metamorphic core complex. Pull a blanket off a dozen sleeping cats, and the cats wake up and arch their backs. In a somewhat analogous manner, metamorphic core complexes—domelike mountains cored with deep basement rock—rose across the West in the early stages of the Basin and Range Orogeny as the crust stretched and the rock above the rising domes slid off.

THE TRAINS creaked south through the City of Rocks, climbing a steady grade to 6,280-foot Pinnacle Pass. Here towering granite ramparts stand sentinel over a gap just wide enough for a single wagon to pass through. Squeezing through one by one, thousands of wagon wheels cut grooves into the solid granite that you can run your hands over today. Stand at Pinnacle Pass today, and you look south out of Idaho into Utah. Had you stood there before February 1848, you would have stood on the edge of Oregon Territory looking south into Mexico. Victory in the Mexican War—ending in the Treaty of Guadalupe Hidalgo in February 1848—brought into U.S. hands the vast territory that today includes all of Utah, Nevada, and California, as well as sections of present Wyoming, Colorado, Texas, New Mexico, and Arizona. (In 1853 the Gadsden Purchase would fill out the boundaries of New Mexico and Arizona to establish the present U.S.-Mexico border.)

Descending from Pinnacle Pass, the trail converged with the Salt Lake Cutoff coming east from Salt Lake City. With that merger, nearly all overland traffic bound for California was now crowded onto one road.[3] Onward they rolled, headed for the Humboldt River 100 miles ahead—but it was too soon to think about that. Just ahead loomed a monstrous Basin and Range ridge, the 50-mile-long Goose Creek Range. There was no way around, and only one way over— 6,960-foot Granite Pass.

The ascent to Granite Pass wasn't bad, just a slow and steady climb. The descent wasn't so easy. The trail plummeted 2,000 vertical feet in just a few miles down to Goose Creek. Bernard Reid thought the descent "the most rugged and difficult for wagons I ever saw," while Elijah Farnham believed that the plunge "would not have been practicable to any but a California emigrant." With practice and skill gained from tricky descents harking back to Ash Hollow on the Great Plains (now more than 800 miles behind), the emigrants slipped and skidded their wagons down the slope. Some "attached small trees to their wagons as a help to let them down easy." Others chained the wheels and steadied the wagons "with long ropes held by the men and thus prevented [the wagons] from sliding against the teams, which were scarcely able to retain a foothold on the rocks."

3. Recall that at Fort Bridger many emigrants chose to leave the original Oregon-California Trail and head for Salt Lake City. From there the Salt Lake Cutoff took them northwest around the lake to rejoin the trail here, a few miles south of City of Rocks. With the merger of the Salt Lake Cutoff, the only California-bound emigrants who were not now on this road were the few who took the Hastings Cutoff. That route passed south of Great Salt Lake and continued west to join the Humboldt River a few miles downstream of present Elko, Nevada.

Safely down at Goose Creek, A. J. McCall[4] paused for a look at his trail-worn companions.

> How times change and men change with them. I look in vain among the ragged grave and bronzed codgers, dragging themselves wearily along, for those dashing, sprightly and gay young fellows, full of song and laughter, whom I saw in the valley of the Blue, on the banks of the Platte, two months ago. Where have they gone and what has become of them? . . . This long and weary migration has sobered these gay sparks and sadly changed their brilliant plumage. . . . As we have no looking glasses, each one can laugh as much as he pleases at the sorry appearance of his neighbor without being cognizant of his own.

Laughter didn't come so easily to others. Few emigrants at this point had not seen significant losses of livestock. "Hardly a day passes when we do not pass more or less dead cattle, horses or mules and sometimes see all of them in one day," Israel Hale wrote. Franklin Langworthy complained that "the effluvia of dead animals fills the surrounding atmosphere." The attrition of livestock meant the abandonment of valuable property. Lorenzo Sawyer cataloged an "immense amount of property strewn all along the road, such as wagons, harness, broken guns, trunks, clothing, &c., abandoned by the owners because their teams have failed."

Food shortages now threatened as well. "Had many applications to-day — as every day we have — for provisions," William Rothwell recorded on July 16, 1850, adding, "Not a few are almost entirely out. We have barely enough, but would rather give in case of suffering than to sell at an exorbitant price as many do." Threats of violence and robbery loomed as provisions dwindled. "This evening a Boston mule company of about seventy is said to be broken up and destitute of provisions," Joseph Middleton wrote on August 26, 1849. "[They are] troubling emigrants to sell provisions to them. They say if people will not sell to them, they will take by force. . . . Such men are to be dreaded and must be sharply watched."

There was nothing to do but press on. Having gained the valley of Goose Creek, everyone now turned southwest, rolling upstream toward a divide that would take them out of Snake River drainage and into Great Basin drainage. Notwithstanding recent hardships, the Goose Creek Valley was a beautiful section of the trail. Franklin Langworthy described it as "a stupendous amphitheater of naked, barren mountains" — and it is no less stupendous today. Tabletop bluffs eroded from pastel stacks

---

4. You may remember forty-niner A. J. McCall from earlier chapters. We left him at Fort Bridger. He took the Salt Lake Cutoff and thus rejoined the main trail south of the City of Rocks. His overland diary is one of the most delightful of emigrant accounts.

of volcanic rock (thick beds of ash spewed from the Snake River Plain calderas) line the valley. Juniper trees—those gnarled icons of the high western desert—cluster on the bluff tops, while sagebrush and blooming yellow rabbit brush cascade down the slopes to the valley floor. Hip-high grass paves the bottomlands. A single gravel road winds along the valley today, reaching out to the widely scattered ranches. Not one square foot of asphalt has yet to touch down in 100 quiet and lonely miles between Goose Creek Valley and the Humboldt River.

The Goose Creek Valley was anything but quiet in gold rush days. An endless stream of creaking wagons and lowing cattle rolled in clouds of dust toward the low divide that separates Goose Creek from the Thousand Springs Valley beyond. Crossing that divide, the emigrants again entered the Great Basin. They would travel across the Great Basin for the next 500 miles.

Thousand Springs Valley served as a fine introduction to the desiccated tramp across the Great Basin. The name conjures up an image of a bucolic valley overflowing with cool, gushing springs. Reality is starkly different. "This valley certainly has the wrong name," William Gordon complained. "It should be the Valley almost destitute of springs, at least so far as my observation extended." Anyone traveling the Thousand Springs Valley today can appreciate Gordon's disappointment. The springs are few and far apart. A few are quite pretty, but most are wretched, cattle-tramped wallows splattered with cow turds and buzzing with flies—a condition that probably mimics their appearance during the heavy traffic of the gold rush years. For the most part, the valley is a parched, dusty, wind-whipped place that only sagebrush could love. "This part of the country is uninviting to the eye of the traveler," Mica Littleton asserted. The dead animals strewn along the trail in emigrant days didn't help the scenery. "The last few days we have seen hundreds of dead oxen, mules and horses," Littleton wrote, adding, "This is truly a long and difficult trip on man and beast."

Some of the springs in the Thousand Springs Valley flowed hot, others cold, and some flowed hot and cold side by side. "This morning we . . . came to springs that were boiling hot," Margaret Frink recorded on July 21, 1850. "Only five feet from them was another as cold as ice. Here were men engaged in washing their clothing. Their position was such that, after washing a garment in the boiling springs, they could take it by the waistband and fling it across into the cold spring, and vice versa, with perfect ease." Byron McKinstry, coming to the same spring[5] a few

---

5. This hot-and-cold spring lies today on the property of the Wine Cup Ranch close by the gravel road that runs along the Thousand Springs Valley. A Bureau of Land Management kiosk marks the site.

10.5 Rock Spring, Thousand Springs Valley, northeastern Nevada. This was the first water the emigrants came to after leaving Goose Creek and crossing into the Great Basin. "Consequently it is generally crowded with footmen, horses, cattle & wagons," Augustus Burbank recorded on August 7, 1849. Rock Spring was (and still is) one of the only pleasing sights in the otherwise parched and dreary Thousand Springs Valley. The spring flows from fractured, tilted layers of volcanic andesite.

weeks later, thought it "would be a glorious place for scalding hogs." With no hogs at hand, he took a bath instead.

> Wishing to bathe and finding one stream too hot & the other too cold, the right temperature was obtained below the junction, where we had fine sport among the shoals of little fishes that would ascend to the junction but we could not drive them into the hot water or we could have cooked a bushel for dinner. Not succeeding in driving our meal into the pot we consoled ourselves by eating currants which grew in abundance in the thicket on the bank.

In the Thousand Springs Valley, the landscape opens up into the vast spaces so typical of the Great Basin. Bare-rock mountains step off to the horizon, their profiles sharp against the sky. Broad basins paved with sagebrush and sparse grass sprawl for miles in between. Juniper trees dust the mountain flanks, and forests of conifers dot the highest ridgelines. You have to walk a long way uphill to find a respectable tree in the Great Basin. To most of us, the concept of "tree line" implies an upper limit to forest growth. It is the elevation (or latitude) *up* to which trees can grow, and above which cold and snow halt their advance. But in the Great Basin, as in much of the arid West, the tree line is more often a lower limit. It is the elevation *down* to which trees can grow, and below which aridity thwarts the germination of seedlings. The mountains of the Great Basin catch more precipitation than do the valleys. The higher you go, the wetter it gets. The ranges thus wear their forests up high, like short green skirts above long legs of naked rock.

Cresting the divide that separates the Thousand Springs Valley from the upper reaches of the Humboldt River, the wagon trains descended a long alluvial plain to Humboldt Wells. Here, near the present town of Wells, Nevada, lay a cluster of springs that emigrants generally regarded as the headwaters of the Humboldt River. The infant Humboldt here begins its 350-mile journey west to the Humboldt Sink.

Gaining the head of the Humboldt River was one of the great milestones of the westward journey. Here was a river that would lead all who followed it nearly to California. Yet mixed feelings came with the first sight of this notorious, maddening, and indispensable stream. While arrival at the Humboldt meant that the wearisome journey was three-quarters over, it also heralded greater hardships to come. "We are now encamped at the headwaters of the dreadful Humboldt," sighed John Hawkins Clark, "of which such hard stories have been circulated on the road."

BASIN, RANGE, basin, range—a visual cadence, clearest when seen from the air. You can see it from the ground too, if you get up high enough. Forty-niner Alonzo Delano saw it as he looked west from Granite Pass in the Goose Creek Range. "It had evidently been the scene of some violent commotion," he wrote of the grand sprawl of mountains before him, "appearing as if there had been a breaking up of the world." It is a perceptive comment, for the Basin and Range Province is indeed a broken-up world—a place where the Earth's crust has stretched by several hundred miles, cracking into hundreds of basins and ranges in the process.

Mountains like those of the Basin and Range don't rise like loaves of bread. They jump up a bit during earthquakes, and then slowly crumble down while stresses rebuild along their flanking faults. Then they jump again—by inches, by feet, by tens of feet even—gaining back in a violent jolt what erosion took away and sometimes more. Erosion, the indefatigable tortoise, is the eventual winner, but the forces of uplift can lead the race for millions of years. And even as old mountains crumble down to plains, new ones are on the rise somewhere else—the shifting regimes of

10.6 Back-to-back earthquakes on December 16, 1954, tore these fault scarps along the eastern flank of the Stillwater Range (upper photograph) and Fairview Peak (lower photograph) in western Nevada. Both mountains grew about seven feet taller and slid north about eleven feet during the quakes. Such incremental uplifts are what made the mountains of the Basin and Range Province. Forty-niner Alonzo Delano recognized this—at least in a general sense—when he described the Basin and Range landscape looking like "there had been a breaking up of the world," and he added, "It is an interesting field for the geologist, as well as for the lover of the works of nature."

the Earth's tectonic plates make sure of that. There has probably never been a time when the Earth did not have mountains, even though any one range makes but a cameo appearance in the epic of time.

We need not worry about erosion winning the race anytime soon in the Basin and Range. The mountains there are leaping up like starving dogs at dangled meat. They are young ranges, bearing the scars of ongoing uplift. Snug against the east base of Nevada's Fairview Peak lies a cliff up to 15 feet high that runs for six miles along the mountain front. It is a fault scarp—the surface break of a rupture that penetrates miles into the crust. It appeared in a virtual instant on December 16, 1954, when a violent earthquake heaved Fairview Peak seven feet higher in little more time than it takes you to read this sentence. The scarp ripped through nearby Highway 50, busting it into a series of asphalt steps. Four and a half minutes later, another quake 20 miles north lifted the Stillwater Range seven feet. There are dozens of similar young scarps throughout the Basin and Range, cutting across the toes of the ranges. Earthquakes let mountains up. They do it by releasing built-up stress along faults. Since the days of the gold rush, nine earthquakes larger than the 1994 Northridge quake (which reduced large sections of the greater Los Angeles area to rubble) have wracked the Great Basin region of the Basin and Range. Each one made a nearby mountain range higher and/or the adjacent valley lower. Hundreds of smaller quakes pop off each year throughout the province. The cumulative work of the quakes has put the ranges, on average, 3,000 to 6,000 feet above the basin floors.

But that's not the half of it. The visible topographic relief of the Basin and Range landscape is but a fraction of the structural relief—the actual distance that the ranges have risen relative to the basins. The basins are not empty holes; they are filled in with thousands of feet of alluvium— the broken debris of the rising ranges. Jackson Hole, the basin east of the Teton Range in westernmost Wyoming, holds 16,000 vertical feet of silt, sand, and gravel. No one forgets the spectacle of the Teton Range towering 6,770 feet over the valley of Jackson Hole (the view is the main attraction of Grand Teton National Park), but the depth of the hole is more spectacular still. If you could scoop all 16,000 feet of dirt out of Jackson Hole and stand in the bottom, you would be staring at a *four-and-a-half-mile rise* from basin to range—more than the uphill trip from the beaches of Chile to the top of Aconcagua, the highest peak in the Western Hemisphere.

THE MOUNTAINS of the Basin and Range did not form the way that most mountains do—by sideways squeezing and thickening of the crust during subduction or continental collision. In other words, they did not form the

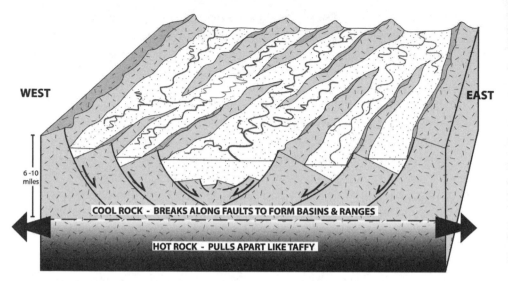

WEST

EAST

6-10 miles

COOL ROCK - BREAKS ALONG FAULTS TO FORM BASINS & RANGES

HOT ROCK - PULLS APART LIKE TAFFY

10.7 East-west stretching of the Earth's crust made the Basin and Range Province. As the crust pulled apart, the cooler rocks in the upper crust broke into separate blocks along large, curving faults. Rocks deeper in the crust, being hotter, flowed apart like taffy, allowing the broken blocks above to rotate. The up-rotated parts of the blocks form mountain ranges, while the down-rotated parts form basins filled with several thousand feet of rock debris eroded from the ranges. The figure shows (very schematically) how west-tilting ranges dominate the western half of the Basin and Range while east-tilting ranges dominate the eastern half, although there are plenty of exceptions to this rule.

way the Appalachians, Andes, Alps, Pyrenees, Caucasus, Zagros, Himalayas, Canadian Rockies, Cascades, Overthrust Belt, or Foreland Ranges did. They formed by the opposite force—stretching.

In the Basin and Range, the Earth's crust has stretched east-west and cracked into dozens of north-south-trending basins and ranges. Cool rock breaks when you stretch it, whereas hot rock pulls apart like pizza dough. In the Basin and Range, the upper six to ten miles of the crust is cool enough that east-west stretching has broken it into north-south-trending faults. These faults curve and flatten out as they go down. The hot, soft rock below, oozing to the east and west, acts like a rug pulled out from under the rigid blocks above, rotating them along the curving faults. The upward-rotated parts of the blocks make the ranges; the downward-rotated parts make the basins. You can see the rotation in the tilted rock layers of the ranges; most of the ranges tilt either east or west. The Basin and Range Province is one of the mostly highly stretched places on Earth. If you add up all the displacements on all the faults that divide the basins from the ranges between Reno and Salt Lake City, you come up with *250 miles* of east-west extension. Given that Reno and Salt Lake City now lie 450 miles apart, that means that east-west stretching has more than doubled the width of the crust. A map of California shows how the coast-line bulges into the Pacific Ocean. The east-west stretching of Nevada and Utah pushed it out there. During Basin and Range stretching, a 400-mile-long block of granite that once lay near Las Vegas was pulled 150 miles west and tilted up into the air. Today we call it the Sierra Nevada.

Those are the basic facts of the Basin and Range—a land stretched east-west to more than twice its original width, and broken up into north-south basins and ranges separated by great faults. This stretching went hand in hand with a broad collapse of the land surface—the singular fact that explains why the Great Basin lies *within* the Basin and Range (figs. 10.2 and 10.3). As the crust stretched and thinned, it sagged in the middle to form a region of inward drainage—the Great Basin. Fossil plants from western Nevada show that the region stood about two miles high before stretching commenced. This high, pre–Basin and Range landscape was, in all likelihood, left over from the Sevier Orogeny, when massive sideways squeezing from the west thickened the crust with folds and thrust faults. About 20 million years ago, the Basin and Range Orogeny started to yank this thick, high crust apart so that it collapsed down to its current one-mile-high average elevation. But what made that happen? And why, given how very stretched and thin it is, does the Basin and Range still stand so high? To get at these puzzles, we need to look for a moment at how the Earth holds up mountains.

An iceberg calving from a tidewater glacier plunges into the ocean and disappears in a great splash. But then it bobs up and adjusts itself to float at a balanced level, with about 15 percent of its total mass exposed above the ocean surface. The 15 percent figure comes from the fact that glacial ice is, on average, about 15 percent less dense than seawater. The Earth's crust makes similar adjustments as it floats in the mantle below—a concept known as isostatic adjustment. The granites, gneisses, and schists that dominate the continental crust are less dense than the peridotite rock that makes up much of the mantle—about 15 percent less dense, in fact, like glacial ice compared to seawater.[6] The crust thus floats buoyantly

6. Here are the actual numbers for readers who want to know. Seawater density averages 1.025 grams per cubic centimeter (gm/cc). The density of ice without air bubbles is about 0.92 gm/cc, but typical glacial ice is a bit less dense—about 0.87 gm/cc—because it holds trapped air. The density of glacial ice compared to seawater is therefore about $(1.025 - 0.87)/1.025 = 0.15$. In other words, glacial ice is about 15 percent less dense than seawater. The continental crust is mostly granite, gneiss, and schist, with an average density of about 2.80 gm/cc. The rock of the upper mantle (mostly peridotite) has a density of about 3.30 gm/cc. The density of continental crust compared to mantle is therefore about $(3.30 - 2.80)/3.30 = 0.15$. In other words, continental rock is about 15 percent less dense than mantle rock—the same difference as glacial ice compared to seawater.

in the mantle in a ratio roughly like that of a floating iceberg. (Although little of mantle is molten, much of it is hot enough to deform and flow slowly, allowing the crust above to make this buoyant adjustment.) Just as thicker ice makes taller icebergs, thicker crust makes taller mountains.

This simple concept—we can call it the iceberg concept—works well for many mountain belts. The iceberg concept holds that every eight miles of crustal thickness will yield roughly one additional mile of elevation. In other words, since continental crust at sea level averages about 20 miles thick, a 28-mile-thick crust will yield about a one-mile rise in elevation, a 36-mile-thick crust will yield a two-mile rise in elevation, and a 44-mile-thick crust will yield a three-mile rise in elevation. Both the central Andes and the Tibetan Plateau average about three miles high. Seismic wave studies show that the crust under those mountain belts ranges 40 to 45 miles thick—about what the iceberg concept predicts.

But not so fast. When geologists in the 1990s completed the first comprehensive seismic surveys of the western United States, they were baffled to discover that the crust in many areas of the West is far *thinner* than expected, given the high elevations. (The crust under the Colorado Rockies, for instance, is only two to three miles thicker than it is under the Great Plains, whereas the iceberg concept predicts that it should be at least eight miles thicker.) Nowhere is the disconnect between crustal thickness and elevation more apparent than in the Basin and Range. Here, where the land averages a mile or more above sea level, the crust is as little as 15 miles thick—*less* than typical sea-level continental crust. The iceberg concept tells us that crust that thin shouldn't be standing one mile high. It should be awash under ocean waves. The basins of the Basin and Range should be flooded like deep Norwegian fjords, with the ranges standing as wave-battered islands in between. As it turns out though, the Earth knows more than one way to hold up a mountain belt. In the Basin and Range—and perhaps in other areas of the West—it is *heat,* not crustal thickness, that makes the region stand high.

The hottest part of the Basin and Range lies below Battle Mountain, a town of 2,800 people and not many fewer billboards straddling Interstate 80 and the Humboldt River in north-central Nevada. The town lies on top of the Battle Mountain High, a 100-mile-wide bull's-eye where heat seeps out of the ground at nearly two times the global average. Throughout the Basin and Range, heat pours out of the crust at rates that average 50 percent higher than typical for the rest of the Earth. This tells us that hot mantle lies close to the surface below the Basin and Range. Since hot rock is inclined to rise buoyantly, it seems likely that the mantle under the Basin and Range is bulging and spreading like a growing mushroom. This hot, rising mantle both stretches the crust and sustains the high elevations. Taking advantage of the thinned and broken crust above, mantle-derived magmas have jetted upward to stain the Basin and Range landscape with innumerable lava flows and volcanic cones.

So, an immense—and ongoing—outpouring of mantle heat appears to have made the Basin and Range and kept it standing high. But what made the heat? For that we turn to the Farallon Plate.

IN CHAPTER 7 we saw how the subduction of the Farallon Plate triggered mountainous upheavals across the West. Flattening through time as it dove under the continent, the Farallon Plate spawned the Nevadan, Sevier, and Laramide orogenies in an eastward-advancing wave of compressive mountain building that climaxed in the uplift of the Foreland Ranges. Meanwhile, the continent grew to the west by hundreds of miles as one terrane after another, cruising in on the Farallon Plate, landed on the west coast to assemble most of what would become California, Oregon, Washington, and British Columbia. Like immense freighters colliding with the dock of North America, the terranes pushed with titanic force as they crunched into the continent, contributing to the compression that squeezed up the mountains far inland. We left the Farallon Plate in chapter 7 lying flat beneath the newborn Foreland Ranges at the end of the Laramide Orogeny 45 million years ago. To understand the Basin and Range—a product of sideways *stretching* instead of compression—we need to look at what happened to the Farallon Plate after that.

The Farallon Plate is largely gone today, snuffed out by North America's westward migration. In the last few tens of millions of years, the North American Plate—migrating west—has overridden a large section of the East Pacific Rise, the spreading mid-ocean ridge from which the Farallon Plate originally grew and slid east. The west edge of the continent first overtopped the spreading ridge about 28 million years ago, meeting the Pacific Plate on the other side. Whereas the Farallon Plate and the North American Plate had been heading in opposite directions (so that one dove beneath the other), the Pacific Plate and the North American Plate were going in nearly the same direction. The slight difference in their relative directions—west for North America, northwest for the Pacific Plate—resulted not in subduction at their touching edges but in side-by-side sliding. A new plate boundary was thus born—a side-by-side sliding boundary represented today by the San Andreas Fault. Farallon Plate subduction ended wherever North America overtopped the East Pacific Rise and made contact with the Pacific Plate on the other side. That region of terminated subduction now extends 2,000 miles from Cape Mendocino in northern California south to Puerto Vallarta, Mexico, at

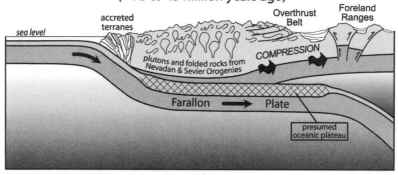

FLAT SUBDUCTION of the LARAMIDE OROGENY
(~ 75 to 45 million years ago)

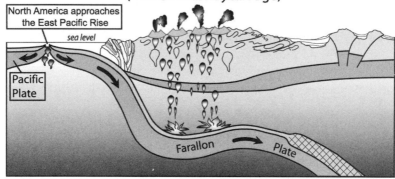

CALDERA ERUPTIONS across the future BASIN & RANGE,
(~ 43-21 million years ago)

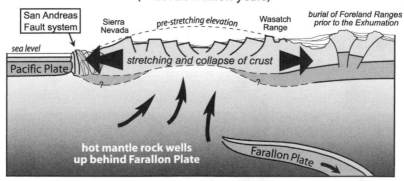

FORMATION of the BASIN & RANGE
(~ last 20 million years)

10.8 The plate tectonic history of the Farallon Plate and western North America, from the Laramide Orogeny through the Basin and Range Orogeny.

the mouth of the Gulf of California.[7] Precisely within those brackets of latitude lies the Basin and Range. That's no coincidence. The termination of Farallon Plate subduction probably made the Basin and Range.

As North America closed like a sliding trapdoor over the Farallon Plate, the plate did not go quietly into the mantle night. It raged and raged, first searing vast tracts of the West with fiery volcanic clouds and then stretching the crust to make the Basin and Range. The plate's death throes began about 43 million years ago, as it started to peel away from the base of the North American Plate to end the flat subduction episode that had hoisted the Foreland Ranges.[8] Restoring itself to normal subduction mode, the Farallon Plate began once again to crank out magma. The result was a maelstrom of volcanism that would rival even that of the later calderas of the Snake River Plain-to-Yellowstone tract. Between 43 and 21 million years ago, calderas opened fire all across what would become the Basin and Range. Incandescent clouds of volcanic ash incinerated the landscape—and every living thing on it—time and again before settling, crackling hot, to weld into layers of volcanic tuff. You can see these tuff layers today stacked up hundreds of feet thick throughout the ranges of the Basin and Range, painting the mountainsides with lovely bands of pink, ochre, and gold.

Meanwhile, North America closed in on the East Pacific Rise and about 28 million years ago began to override it, smothering ever-larger sections of the Farallon Plate in the process. With no more East Pacific Rise to continue spreading and cranking out more ocean floor, the Farallon Plate

7. North of Cape Mendocino and south of Puerto Vallarta, two remnants of the Farallon Plate are still subducting under North America. From Cape Mendocino north past Vancouver Island, a fragment of the Farallon Plate called the Juan de Fuca Plate dives under the continent at the Cascadia Trench. Its volcanic offspring form the Cascade Range, including Mount Shasta, Mount Hood, Mount Saint Helens, Mount Rainer, and about a dozen other volcanic peaks. (Some geologists consider Mount Rainer to be the most dangerous volcano on the planet. It perches like a magma time bomb on the edge of the heavily populated Seattle-Tacoma area.) South of Puerto Vallarta, a larger Farallon Plate remnant called the Cocos Plate dives under Mexico and Central America at the Central America Trench. These two trenches—the Cascadia Trench to the north and the Central America Trench to the south—were once part of the continuous Farallon Trench, running unbroken from Canada to Mexico, before North America overrode the East Pacific Rise.

8. You may recall from chapter 7 that, according to current theory, subduction of a buoyant oceanic plateau may have floated the Farallon Plate up so that it scraped flat against the bottom of North America to raise the Foreland Ranges. Presumably, ocean floor of more normal density followed the buoyant plateau down the trench, pulling the Farallon Plate back down to a more normal subduction angle. Fig. 10.8 illustrates the change.

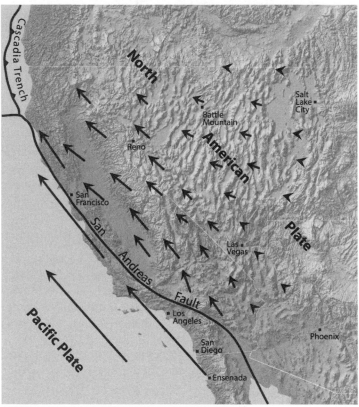

Rate and direction of movement relative to the interior of North America

→ 0.5 inches per year

———→ 1.0 inches per year

————————→ 2.0 inches per year

10.9 The stretching of the Basin and Range continues today. The figure shows how fast different parts of the southwestern United States are moving relative to the interior of North America (based on GPS satellite measurements). The length of the arrows relates to the rate of movement. Regions west of the San Andreas Fault (which include Baja California as well as the San Diego and Los Angeles areas) are all part of the Pacific Plate and are moving northwest at about two inches per year relative to the interior of North America. Notice that much of California and Nevada east of the San Andreas are also moving northwest with the Pacific Plate—even though these areas are traditionally considered part of the North American Plate. Look up the San Andreas Fault in any geology textbook, and odds are you will see it described as "the boundary between the Pacific Plate and the North American Plate." But reality is not so simple. Vast areas of the southwestern United States are being dragged northwest with the Pacific Plate. In the future, if California and western Nevada continue to tear away from the rest of North America, an ocean basin may open up through Nevada.

was cut off from its source. It sank like a scuttled ship into the mantle below North America. You can see it there today. It shows up on seismic images as a 2,300-mile-long slab of rock, cooler than the surrounding mantle, angling down to the east. Its leading edge nearly touches the Earth's core 1,800 miles below Bermuda, while its trailing edge hovers a few hundred miles below the western Great Plains. To the west, under the Basin and Range, seismic imaging reveals a large pocket of hot mantle rock that appears to be welling up through the gap opened by the departed Farallon Plate. Many geologists, in a grand but reasonable arm wave, suggest that that hot mantle—moving up behind the Farallon Plate as the plate exited stage right—explains the heat and crustal stretching of the Basin and Range (fig. 10.8). No longer squeezed sideways by Farallon subduction, the crust—already thick from the Sevier Orogeny—was inclined to relax and spread out east-west anyway. The hot mantle that welled up underneath encouraged it to do that. The timing fits. Studies of Basin and Range faults show that crustal stretching began spottily about 36 million years ago, near the end of Farallon subduction. But the stretching didn't really kick into high gear until 20 million years ago—about when the caldera eruptions ended, signaling that the Farallon Plate had left the picture for good.

POUND A NAIL through the North American Plate in central Utah, and everything to the east would stop. But California would still move west—and so would everything in between, only less so. The stretching of the Basin and Range is a work in progress. The evidence spreads across the broken land as fresh fault scarps cutting across the feet of the mountains, as innumerable steam vents and hot springs, and as a never-ending drum roll of tiny earthquakes punctuated by occasional bigger jolts. Global Positioning System satellites today measure how fast the crust is stretching—and they reveal a new twist. The Pacific Plate appears to be directing the present stretching of the Basin and Range. Bound by friction to the Pacific Plate along the San Andreas Fault, much of California and the western Basin and Range is pulling away from the rest of North America, at speeds approaching a half inch per year (fig. 10.9). You wonder—does the Basin and Range belong to the North American Plate or the Pacific Plate? Perhaps to both. Only geologic time will reveal its allegiance, perhaps when an ocean basin opens up through Nevada—possibly right through Battle Mountain, where the crust is thinnest and weakest. Seas will then divide the desert, just as the opening of the Red Sea has split Arabia from Africa, and as Baja California has torn away from Mexico. Beachfront property in central Nevada is well within the Earth's bag of plate tectonic tricks.

MOST MISERABLE RIVER

> Perhaps the Devil himself having cast his eyes over the world con-
> cluded to try his hand at making a river. He made it in the night and
> laid it down so crooked and ragged, that just at break of day when
> he stopped to look back at it, he got ashamed of himself and ran it
> into the ground.
>
> JAMES EVANS, 1850, along the Humboldt River

It is one of the most important rivers in American history. It is not mighty
or majestic, nor is it imbued with noble folklore like the Mississippi or
the Missouri. The only poetry it ever inspired oozes with loathing. It is a
forlorn little stream in northern Nevada, snaking west through the poor
grass of its valley. Few people today outside of Nevada could point to it
on a map. Yet those who came to know the Humboldt River would never
forget it.

Adison Crane was one. On August 14, 1852, after sixteen days following its meandering 350-mile path west, he waved good-bye to the river.

> Farewell to thee! thou Stinking turbid stream
> Amid whose waters frogs and Serpents gleam
> Thou putrid mass of filth farewell forever.
> For here again I'll tempt my fortunes never
>
> For Sixteen gloomy, sad & weary days,
> 'Mid burning Sands and Sols more burning rays
> I've wandered on thy grassless Sagy plain
> And took thy putrid current to my veins
> Drank by compulsion of the brothy mass
> That in these deserts must for water pass

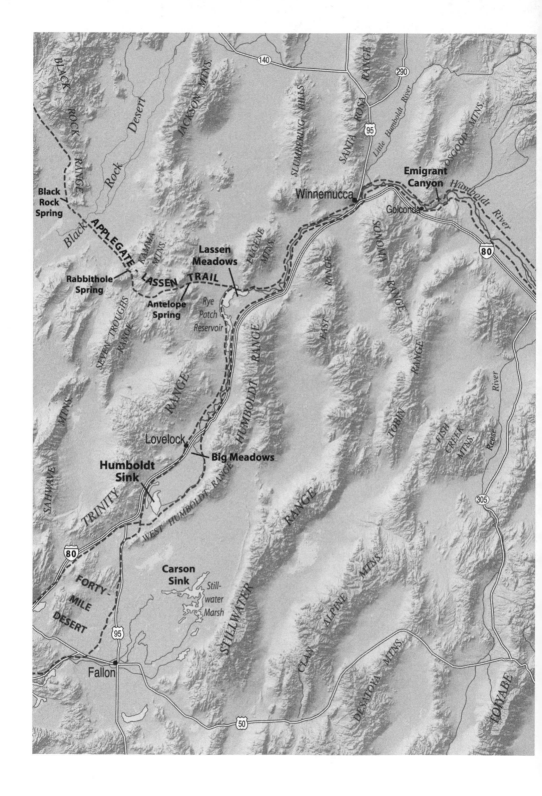

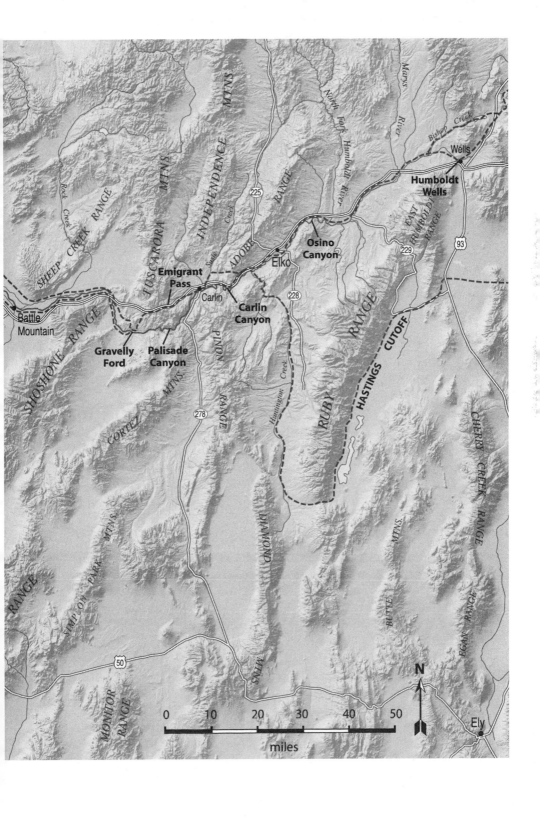

11.2 The Humboldt is a skinny stream bisecting an arid land. Its water comes mostly from winter snow caught by the mountain peaks of the Great Basin, such as the Humboldt Range in the left distance. The view is southwest from the Tungsten Road bridge a few miles west of I-80 exit 149.

Dr. Horace Belknap also felt moved to verse by the Humboldt's muses. In 1850 he announced to the river:

> Meanest and muddiest, filthiest stream, most cordially I hate you;
> Meaner and muddier still you seem since the first day I met you.
> Your namesake better was no doubt, a truth, the scriptures tell.
> Her seven devils were cast out, but yours are in you still.

The Humboldt River flows murky and green. Ropes of algae wave in its sluggish current. Among American rivers, it's a runt—rarely more than 40 feet wide and thigh-deep. Yet it is the only permanent river crossing the Great Basin. Drink its water and you taste a bit of emigrant life on

11.1 (*previous pages*) The Humboldt River was the emigrant's lifeline across the arid Great Basin. The river runs 350 miles west-southwest across Nevada from its source at Humboldt Wells to the Humboldt Sink. It drops 1,730 feet over that distance, for an average gradient of five feet per mile. Interstate 80 parallels the river and the emigrant trail today.

the California Trail. It's not pleasant. After a fresh rain, or in the upper reaches of its valley, the river is palatable. Most of the time, though, it's worse than awful—distinctly salty, leaving your mouth with a slippery, soapy aftertaste. The surface is often dotted with clusters of bubbles, like seepage from an old Laundromat. Its banks are impossibly muddy. Just to reach the water often means floundering through yards of muck—slurping, fetid muck served up scalding hot under the Great Basin sun. Leander Loomis summed up the prevailing judgment when he wrote, "We have seen more suffering and passed through more hardships since we struck . . . the . . . Humboldt, than in all the rest of our journey."

No one loved the Humboldt River, but no one could leave it either. There was no better way west. There were other routes across the Great Basin. There were trails that passed through present New Mexico and Arizona, and there was the Old Spanish Trail from Salt Lake City to the pueblo of Los Angeles. But these were primarily winter roads, horrendously hot and dry in summer. The Humboldt route was the only road that could support large numbers of people and livestock through the heat and aridity of a Great Basin summer. It was also the most direct route from South Pass to California. Meandering and muddy, with foul water and often poor grass, the Humboldt nonetheless formed a 350-mile-long

11.3 The Humboldt River at Elko, Nevada. The steam comes from a hot spring on the riverbank. This spring flowed more powerfully in emigrant days and was a famous landmark. Forty-niner Wakeman Bryarly described the water as "boiling hot & sent off steam & heat from its surface, which [was] as hot as the scape pipe of an engine. Even after it ran into the river, it foamed & hissed as cold water poured into hot and sent off steam for 80 yds. below, and extended half over the river." Many emigrants bathed and washed their clothes here in the spring-warmed river. The ledge and the lumpy, white slope consist of travertine deposits precipitated by the hot water.

life-sustaining corridor through some of the most "God-forsaken, barren and desolate" (in A. J. McCall's opinion) land in North America. More than a quarter million emigrants funneled west along the river from 1841 to 1869, with the heaviest traffic passing through during the gold rush.

NATIVE PEOPLES eked out a living along its banks for thousands of years before white explorers, but Peter Skene Ogden of the Hudson's Bay Company is credited with discovering the Humboldt River in 1828. He called it the Unknown River since his explorations were limited and did not extend to the river's end. Fur trappers after Ogden called it variously Ogden's River, St. Mary's River, or most commonly Mary's River, Mary supposedly being the wife of one of Ogden's trappers. John C. Frémont bestowed the name Humboldt on the river while on his 1845 expedition, to honor German geographer Baron Alexander von Humboldt. Frémont's fame trumped earlier names, and the river became generally known thereafter as the Humboldt, although some emigrants continued to call it Mary's River.

Scatter a few dozen stubby pencils onto a table. Turn each one in place so it points generally north or south. These are the mountain ranges of the Great Basin. Tempt an ant to find a path westward through the pencils. The ant wanders this way and that, finding a route around the ends of the pencils. This is the meandering westerly route of the Humboldt River, nosing its way west around the ends of the ranges for 350 miles before pooling up in the Humboldt Sink, about 20 miles south of present Lovelock, Nevada. There it dies, swallowed up by thirsty ground and dry air.

The idea that a river could simply end in the desert was an astonishing notion for many emigrants. Hailing mostly from the rainy East and Midwest, these people knew how a proper river should behave. A river got bigger downstream, swelled by the contributions of its tributaries. The Humboldt does the opposite. The river flows west into progressively hotter, drier country. As its tributaries dry up, and as the ground and air continually rob it of water, the river gradually shrinks and becomes more murky and saline. "The stream," Lewis Beers observed, "begins to grow smaller as fast as we descend . . . towards its mouth or rather towards the place where it runs itself into the ground."

For most of its route west, the wandering Humboldt passes around the ends of mountain ranges. But where there is no break in mountain barriers, the river cuts right through, carving narrow canyons that were sometimes impassible for wagons. In these places, the emigrants had to leave the river and climb the mountains to bypass the canyons. It was unnerving to look down on the Humboldt from these high, distant vistas. The thin river seemed nearly swallowed up by rock. "On whichever side you cast your eyes beyond this small valley [of the Humboldt]," Israel Lord wrote from Emigrant Pass in the Tuscarora Mountains, "you behold nothing but naked barren mountains, and mountains beyond and on the top of mountains, and all void or nearly so of vegetation of any kind." John Hawkins Clark didn't much care for the view either.

> Our road this afternoon [up to Emigrant Pass] is up a steep mountainside seven miles long; the steepest, roughest, most desolate road that can be imagined. The mountains that border this valley . . . have a decrepit and worn-out look. . . . It makes a man lonesome and homesick to contemplate their forlorn, deserted and uncanny appearance. Stunted and scattered cedar trees, broken down by the snows and wild winds of the winter season, gives them a sort of ghost-like appearance that makes one shudder to behold.

Clark felt little relief upon reaching the crest of the mountains at Emigrant Pass. "The scenery at this place is wild, desolate and forbidding, without a spark of romance to enliven our spirits," he wrote. "No wood, no grass, and but a scant supply of water; all is rock, rock, rock, as bare of vegetation as a sterile rock can be."

The Humboldt cuts four deep canyons through mountains on its way west: Osino Canyon, Carlin Canyon, Palisade Canyon, and Emigrant Canyon (see fig. 11.1). Here, as in the Wyoming Rockies, the Humboldt confronts us with a riverine puzzle. How can a skinny river cut deep canyons through hard-rock mountains?

In chapter 4 we saw how, during the Exhumation, the rivers of the Rockies cut their range-crossing canyons via superposition, slicing down through once-buried ranges. But superposition won't work as an explanation for the Humboldt's canyons. Superposition requires that mountains become buried under layers of sediment after forming, so that rivers establish paths *above* them and then cut down. The ranges of the Basin and Range are much younger than the Rockies and have never been buried. More likely, the canyon-cutting process along the Humboldt involves antecedence, another way that rivers can cut though mountain ranges. Both theories—superposition and antecedence—date back to John Wesley Powell: soldier, explorer, and geologist extraordinaire of the late nineteenth century.

John Wesley Powell lost his right arm in 1862 fighting for the Union at the Battle of Shiloh. After the Civil War, he set out west to explore, map, and study geology. In 1869 he floated with ten companions in a handful of rowboats down the then-unexplored Green and Colorado rivers—an epic journey that made him an instant national hero. Powell did more with one arm in a lifetime than most men could ever hope to do with two, mapping thousands of miles of sprawling canyon wilderness and publishing groundbreaking books and papers on the geology and geography of the West. On his travels, Powell was struck by the power of diminutive western rivers to cut huge canyons: Lodore, Desolation, Stillwater, Cataract, and Glen canyons, and—the grandest of all—the Grand Canyon, where the Colorado River has sliced down through more than one vertical mile of rock. Powell proposed antecedence and superposition as alternative theories for how rivers could cut such canyons. We saw superposition at work in the Rockies. Antecedence works like this: Imagine, said Powell, rivers flowing across a land surface before there were any mountains. As the mountains rose, the rivers would cut down through the rising rock like a stationary sawmill blade cutting through a rising log. In this theory, the routes of the rivers predate the uplift of the mountains; that is, the rivers are antecedent.

11.4 Entrenched meanders of the Humboldt River. The upper image looks south across the huge meander bend in Carlin Canyon, where the Humboldt cuts through the Adobe Range a few miles east of Carlin, Nevada. Interstate 80, in the right distance, bypasses the bend through a tunnel. The lower image is an aerial photograph of entrenched meanders in Palisade Canyon, where the Humboldt cuts through the Tuscarora Mountains. The parallel lines that shortcut the bends are railroad tracks.

In two of its biggest range-crossing canyons—Carlin and Palisade—the Humboldt River cuts entrenched meanders, where the sheer walls of the canyons follow the hairpin loops of the river right through the mountains. Since rivers normally develop large meander loops only in wide-open valleys (where the rivers have room to wander), the existence of *entrenched* meanders in the canyons tells us that the meanders predate the cutting of the canyons. Picture the ancestral Humboldt meandering west in great loops across a relatively flat pre–Basin and Range landscape. As the crust stretched and broke up during the Basin and Range Orogeny, the river maintained its meandering path across some of the rising ranges. It sliced down apace with uplift, locking the river into a meandering path right through the ranges—at least in some instances. In other places, the ranges may have risen too fast for the river to keep up. The growing mountains shunted the river aside, sending it onto routes around their ends instead of through their middles. Thus we have today's Humboldt, cutting entrenched meanders though the mountains in some places, passing around the ends of the mountains in others.[1]

IF THE OCCASIONAL canyon-forced detours away from the Humboldt made the emigrants better appreciate the river, few noted it. Travel along the river presented it own miseries.

Rivers are never truly fresh because they dissolve salts and minerals out of rock and soil as they flow. Add intense evaporation and take away freshening tributaries, and you have the repellent alkaline soup of the Humboldt. Forty-niner Elisha Perkins pronounced it "barely drinkable from saline and sulfurous impregnation & having a milky color. I think Baron Humboldt would feel but little honored by his name being affixed to a stream of so little pretension." Henry Sterling Bloom described the water as "detestable; it is fairly black and thick with mud and filth," but he added gamely, "there is one advantage one has in using it—it helps to thicken the soup which would be rather thin without it."

The Humboldt grows fouler downstream as it picks up more dissolved substances and evaporation concentrates the salts. "We had not traveled fifty miles down the stream before we found the water gradually becoming brackish and discolored from the salt and alkali of the soil," Margaret Frink wrote. By the time the water pooled up at the Humboldt Sink, it was foul beyond belief. "One can get an idea of how it tastes," Heinrich Lienhard explained, "by making a strong solution of tepid water and bitter salts and adding several rotten eggs." Facing the nearly waterless Forty-Mile Desert crossing ahead, he added, "Only dire thirst and the knowledge that one would have to walk forty miles before coming to real water could force anyone to take a drink of this diabolic liquid."

The Humboldt's crookedness and muddy bottomlands added to the misery. For many miles along its valley, the river meanders in huge loops, its channel often bending back on itself and even cutting itself off. Wakeman Bryarly described it as "so very crooked in its whole course that I believe it impossible for one to make a *chalk mark* as much so. It is a *dirty, muddy, sluggish, indolent stream* [original italics], with but little grass at the

---

1. This explanation is consistent with observations, but that doesn't necessarily make it right. Geologic theories—like the landscapes they attempt to explain—evolve through time, and future research may reveal a different story of the Humboldt canyons.

best of times." The river's meandering creates a bottomland of countless stagnant swampy pools bordering the main channel. These mires, called sloughs, had to be crossed to reach the water in the main channel—an unpleasant and infuriating task because of the acres of colossal muck. "Country all around to world's end dry as powder, no vegetation, except on the river, and here a complete quagmire," Henry Wellencamp complained. John Birney Hill asserted that a "man will mire down & not half try." If the muddy bottomlands blocked access to the main river channel—or if a dry year had mostly dried up the channel—the emigrants had to make do with the foul water of the sloughs. "Had very bad water, had to use water out of the slough, which was covered with a slimy looking moss," William Gordon wrote one August morning in 1850. But he added brightly, "We had another fine mess of frogs this morning for breakfast."

Livestock, desperately thirsty from hauling heavy wagons through the summer heat, had to be stopped from charging headlong for the river, where they might mire down in the sloughs and drown. Instead, emigrants often had to fetch bucket after bucket of water and armloads of cut grass for the animals. "Had to mow grass for our cattle and horses at noon today on the account of the sloughs being so miry," William Gordon wrote, adding, "The sloughs and swamps have been very plentiful for the last few days, and we have had to be cautious about letting our stock go into them on account of their miry condition." The animals occasionally broke free and bolted for the river anyway, where many bogged down in the muck and drowned. This led to some revolting experiences, such as Gilbert Cole's in 1852.

> For about ten days the only water we had was obtained from the pools by which we would camp. These pools were stagnant and their edges invariably lined with dead cattle that had died while trying to get a drink. Selecting a carcass that was solid enough to hold us up, we would walk into the pool on it, taking a blanket with us, which we would wash around and get as full of water as it would hold, then carry it ashore, two men, one holding each end, would twist the filthy water out into a pan, which in turn would be emptied into our canteens, to last until our next camping place.

Looking out across the drowned and decaying corpses along the river, Eleazer Stillman Ingalls pronounced the Humboldt a "burying ground for horses and oxen. The river is nothing but animal broth, seasoned with alkali and salt." The result, as Franklin Langworthy explained, was that "We frequently take our meals amidst the effluvia of a hundred putrescent carcasses."

Besides being a killing field for livestock, the stagnant Humboldt sloughs were fertile breeding grounds for mosquitoes. Edward Harrow on August 5, 1849, recorded the monotonous daily cycle of Humboldt suffering—heat by day, mosquitoes by night.

> The scorching dry heat has left a warm night. . . . Rain seldom falls here, but the yawning ground plainly tells that the scorching sun and whirlwind of blinding dust are always here. Mosquitoes too have again become demons of trouble. Laid myself down, but sleep I could not, and rose at 3 after passing a sleepless night, to commence again the toils and troubles of another day.

Likewise, here is John F. Riker on July 12, 1852:

> Through the day we suffer from the extreme heat, and at night it is almost impossible to either sleep or rest on account of the legions of mosquitoes that infest this region. While some of them thrust their long bills into our bodies like hungry tigers, those that are forced by their superiors to wait their turn, keep up a constant singing and yelping, as though they were at a feast and at last sure of one good meal. To escape their attacks we are compelled to cover entirely over with blankets, baggage, or whatever else we can lay our hands upon; in this manner we get a little rest, and in truth it is but little.

An emigrant with his water bucket—wading through the hot mud and filthy water of a Humboldt slough while oxen carcasses oozed nearby and armies of mosquitoes gathered in the thickets for the coming night— would doubtless have taken little comfort in knowing that the Humboldt is a fairly typical river in its tendency to meander and make sloughs.

Nearly all river channels, if they have room to do so, will meander. Meandering results from differences in water velocity around bends in the channel. As flowing water heads into a bend, most of the water naturally flows to the outside of the bend, pushed there by the same outward-directed momentum you feel in a car going around a curve. The water cuts into the outer part of the bend, eroding the riverbank. On the inside of the bend, the water flows more slowly and drops some of its load of sand and silt. Over time the combination of erosion on the outside of the bend and sediment buildup on the inside of the bend increases the crookedness of the channel.

As the channel evolves into ever more extreme loops, eventually two separate bends may approach one another and join. When this occurs, the river takes the shortcut and establishes a new channel, bypassing the cutoff loop. These abandoned channels, which may be miles long on large rivers, record the curve of the channel like a letter C or U. Geologists call such abandoned channels oxbows because their curves are reminiscent

11.5 An abandoned channel of the Humboldt River in Emigrant Canyon, a few miles east of present Golconda. Such abandoned channels—which the emigrants called sloughs—form innumerable swampy, mosquito-infested depressions along the Humboldt Valley. The emigrants usually traveled well away from river to avoid these swampy bottomlands. In this area, however, the narrowing of the bottomlands through the canyon forced the emigrants to cut across the sloughs.

of the U-shaped pieces of wood that fasten the yokes onto the necks of oxen (see fig. 2.3). The emigrants—despite handling real oxbows every day—didn't use that term. They called the abandoned channels sloughs.

When they first form, oxbows contain standing water—a haven for mosquito larvae. Over time, with no moving water to keep them scoured, these stagnant ponds fill in to become low, swampy depressions. Emigrants along the Humboldt saw oxbows at all stages, from fresh ones holding several feet of water to ones that had progressed to the swampy stage. You can see the same thing along the river today. And if you want a "period rush," as history buffs call it—meaning that you want to transcend time and touch the past in a personal way—then wait for dusk on a summer evening along the banks of the Humboldt River. As the sun slides below the horizon, the keening mosquito hordes emerge from the thickets, proboscises armed and ready. That's when any spark of romance that you might still feel about the westward journey winks out, and you feel only profound gratitude for living in an age of sealed windows and insect repellant.

DOWN THE valley they rolled, growing more impatient by the day. Bare-rock mountains loomed up all around, different from ones already passed and yet the same. "The eye tires & mind wearies of this tasteless

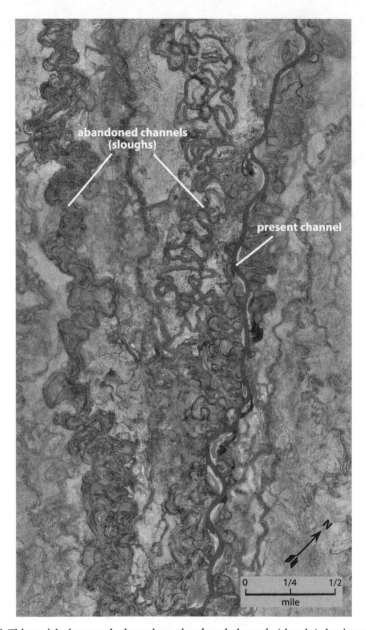

11.6 This aerial photograph shows how abandoned channels (sloughs) dominate the bottomlands of the Humboldt Valley. It was virtually impossible to drive wagons or cattle anywhere near the main channel because of the miry sloughs. The sloughs made the simple task of fetching water from the main channel infuriating, messy, and even dangerous. The area is a few miles northwest of the town of Battle Mountain.

monotony of scenery. The deserts of Arabia could not be more sterile nor more unpleasing to the sight than the country bordering this river," Bennett Clark groaned. Hot days piled up into hotter weeks, and the river grew fouler.

In the valley of the Humboldt, the emigrants met the so-called "Digger" Indians, named for their habit of digging for roots, grubs, and burrowing rodents. There were actually two distinct groups of native peoples along the Humboldt Valley—the Western Shoshone along the eastern and central portions of the valley, and the Northern Paiute along the western portion (west of today's town of Winnemucca, Nevada). Few emigrants made the distinction, though. To most of them, the natives of the Humboldt Valley were simply the Diggers—shriveled and hapless specimens of humanity at best, dangerous thieves and murderers at worst.

While there are reports of friendly exchanges, for the most part the relationship between emigrants and Humboldt Valley natives was one of unremitting animosity. The emigrants, with supplies running low, had little to offer the natives for trade. And the natives—perhaps the most impoverished Indians in North America—had even less to offer in return.

> They were absolutely naked, poor and hungry, and quite in keeping with the character of the country. The average stature of these Root Diggers was not to exceed five feet, and their weight seventy to ninety pounds. Their faces were pinched and careworn, while the most abject misery seemed stamped on every feature, and we looked upon them as types of humanity in its lowest form. (Reuben Cole Shaw, 1849)

Nowhere in emigrant accounts, to my knowledge, is there the slightest acknowledgment of what a stunning feat it was for native peoples to survive on the meager repasts of the Humboldt Valley. The emigrants were at best bemused by the Indians and their starvation diet.

> In the afternoon we met four Digger Indians. . . . Each had suspended from a girdle at his waist a number of ground rats. . . . These were already cooked without disturbance to either tails or entrails. . . . One of the four . . . approached as we rode along. . . . Holding a roasted rat at arm's length, he brought it almost within smelling distance, but our firmness was unabated. He made signs with his teeth and tongue, that it was delicious, but we yielded not a jot. (William G. Johnston, 1849)

"Roots, seeds, and grass, every vegetable that affords any nourishment, and every living animal thing, insect or worm, they eat," John C. Frémont reported during his pioneering 1843 expedition. "Nearly approaching to the lower animal creation, their sole employment is to obtain food; and they are constantly occupied in struggling to support existence."

To a people ruled by hunger—a people grateful to find enough grubs, seeds, and rats to see another sunrise—emigrant cattle must have seemed like a steak buffet sent from heaven. Only the obnoxious invader—the sunburned transient with a gun—stood between the natives and endless beef dinners. But you don't survive for millennia in the Humboldt Valley without being sly and resolute. The Indians found ways. Their favorite trick was to sneak around emigrant camps at night and shoot arrows into the oxen. Pulling up camp the next morning, the emigrants had no choice but to leave their dying animals behind—whereupon the natives descended faster than vultures.

> These Diggers are a small Indian or rather short and have very few guns but are armed with bows and arrows. They are seldom seen on or near the road but keep themselves concealed during the day, and in the night they leave their ambush and sally forth in search of plunder. It has been their custom to cripple stock in such a way that it would become useless to the owner and they would leave it, when the Indians would return and carry off the meat. (Israel Hale, August 15, 1849)

Another native trick was to sneak into the camps at night and quietly lead the cattle away for bloody slaughter.

> We fell in company with a Missouri train, who the night before had five head of cattle stolen. . . . The company followed their tracks twenty-five miles, when they found them with all the cattle slaughtered, and preparations for a grand feast going on. The Indians, however, did not wait to welcome their unexpected guests, but fled to save their own bacon, for the men would have most assuredly shot them had they remained. (Alonzo Delano, August 6, 1849)

Murder and brutality took place on both sides of the cultural divide. Humboldt natives killed and mutilated emigrants . . .

> The emigrant had been shot through the heart with a ball; an arrow was sticking in his skull. His bowels had been torn out, and the heart carried away. We begin to think that these Diggers are somewhat dangerous neighbors. (Franklin Langworthy, September 24, 1850)

. . . and emigrants returned the favor . . .

> Ordered the men to shoot every Indian they may see, until we come into a section where they will come into camp in the day time. These fellows are really the Arabs of America. (Israel Lord, September 2, 1849)

BETWEEN NATIVE depredations, the miry Humboldt sloughs, and the desert that lay beyond the Humboldt Sink downstream, it was a minor miracle for anything on four hooves to make it from the head of the Humboldt River to the Sierra Nevada. Emigrants who lost their livestock, and who could not buy or beg replacement animals, often had to abandon their wagons—which meant leaving behind property, food, and shelter. Those so stranded (if they could not prevail upon others to take them in) had to walk to California carrying what they could on their backs. "I see at least a dozen men daily with their packs on their backs and the number is constantly increasing," Byron McKinstry observed. Since it was impossible for anyone to carry enough food to last even a small portion of the remaining journey, these so-called "foot packers" were often desperate and starving.

> We have been no little troubled today with starving emigrants begging for provisions. There is some that is tetotelly out; others say they have not tasted meat for fifteen days; some complaining of being verry weak with hunger; some of them that is suffering or complains of suffering we divide with but if we was to listen to all we would not have a mouthfull of provisions in two days. God only [k]nows what they are to do. (Thomas Christy, July 16, 1850)

"Often, almost daily," Eleazer Stillman Ingalls explained, "some poor starved fellow comes up to the wagon and pray us in God's name to give or sell him a crust of bread, some of them asserting that they have eaten no food for two or even three days." Hardened by their own worries about getting through, emigrants with food remaining were often reluctant to help those in need. "I presume that twenty men per day apply to us for food," Byron McKinstry wrote on August 18, 1850. "They must eat horses and carrion I am afraid. There must be a great deal of suffering on the road yet." Carrion it would be, for many. Eleazer Stillman Ingalls "noticed several dead horses, mules, and oxen by the roadside, that had their hams cut out to eat by the starving wretches along the road."

Tearing a page from the natives' cookbook, some starving emigrants started digging. "We came on the bank of the river and all go to catching what we term ground rats; they are little mouse colored animals that live in the ground, not quite as large as a small rat. . . . We caught enough of these little animals to make a good meal for our supper and had some left over for breakfast." Carrion, ground rats, roots, and tubers—it was hard to do better along the parsimonious Humboldt. "Here, on the Humboldt," Horace Greeley wrote on his 1859 stagecoach journey, "famine sits enthroned, and waves his scepter over a dominion expressly made for him."

SOME 260 miles downstream from Humboldt Wells, the emigrants came to where the river makes a right-angle turn south for its final 90-mile

Dark near and harsh outs.

M. De Mott
of
Col. Ohio
died
Sept. 18, 1849
Abcdeef
of
camp Fever

11.7 Emigrants who left the Humboldt River on the Applegate-Lassen Trail could count on water from just a few widely scattered springs for the first 90 miles. Rabbithole Spring—sketched here by J. Goldsborough Bruff on September 20, 1849—was one of two springs between the river and Black Rock Springs on the far side of the Black Rock Desert (see fig. 11.1). Emigrants dug shallow wells here to increase the meager water supply. One of Bruff's trail companions described the scene on the day Bruff made his sketch. "The whole environment as far as the eye could reach was simply an abomination of desolation. . . . More than half of the wells were unavailable as they were filled with the carcasses of cattle which had perished in trying to get water. To add to the natural horrors of the scene, about the wells were scattered the bodies of cattle, horses, and mules which had died here from overwork, hunger, and thirst; broken and abandoned wagons, boxes, bundles of clothing, guns, harness, or yokes, anything and everything that the emigrant had outfitted with." Reproduced by permission of the Huntington Library, San Marino, California.

run to extinction at the Humboldt Sink. Here the river splits into numerous rivulets to nourish a vast, grassy plain known as Lassen Meadows. The trail splits as well, presenting everyone with a critical decision. They could stick with the Humboldt for 90 more miles, down to the sink, and then break for the Sierra Nevada across the Forty-Mile Desert. Or they could leave the river here, at Lassen Meadows, and head west along the Applegate-Lassen Trail. This route, blazed in 1846 by guide Jesse Applegate as an alternative southern road to Oregon, also took you to the California gold fields—but by a long, roundabout way through the northern Sierra Nevada. It took one additional month, on average, to get to the gold fields by this northern route compared to the Truckee or Carson routes across the Forty-Mile Desert, but the mountain passes this way were much lower. To get to California by this route, you followed the

Applegate Trail northwest almost to the Oregon border and then turned southwest onto a trail to the Sacramento Valley pioneered by rancher Peter Lassen in 1848. In 1851 William Nobles blazed a cutoff that shortened the distance to California, and this Nobles Trail became a popular option thereafter. But either way, there was no avoiding a bleak and nearly waterless trek across the Black Rock Desert.[2]

Whether you left the Humboldt River here at Lassen Meadows or downstream at the Humboldt Sink, you stepped off into a desolate alkali wasteland. Either way lurked thirst, hunger, abandonment, and death. Everyone hated the Humboldt, but it did promise 90 more miles of water—even if it was Humboldt water. Most emigrants gritted their teeth and stayed with the river.

> Our great want, now is: water! water!! water!!! good spring water, good well water, good snow water, good river water. Our dreams are of water, clear and cold, spouting from the earth like a geyser; the mountain streams that come tumbling over the great boulders . . . are a part of our midnight visions. (John Hawkins Clark, August 13, 1852)

Onward they rolled—south now—toward the Humboldt Sink. Several days' travel along the shrinking river brought them to the area of present Lovelock, Nevada—an agricultural town where today the river is siphoned onto thousands of acres of irrigated fields. The sink lies 20 miles beyond. In emigrant days here, the sullen river—as if repentant on its deathbed—expired in a blaze of glory. No longer confined to a channel, it spread its final dribbles across the flat valley to nourish a vast marsh of luxuriant grass. The emigrants called it Big Meadows or Great Meadows. "It would almost seem that these extensive meadows were placed here expressly to supply the means of traversing this desert country," Lorenzo Sawyer mused, adding, "They are precisely at the point where they are most needed." Emigrants often camped for several days at Big Meadows while their cattle recuperated and cropped the lush fare. There was work to be done at Big Meadows. Practically no grass lay on the Forty-Mile Desert ahead, and so the scythes came out of the wagons and the men starting swinging.

2. These northern routes to California can best be seen on the chapter 13 map. Our arrival at the Applegate-Lassen Trail fork means that we will leave three forty-niners whose accounts have contributed much to this story so far. J. Goldsborough Bruff, Alonzo Delano, and William Swain all tried their luck with the Applegate-Lassen route. Each made it to California, with much attendant suffering. We'll meet them again on the other side of the Sierra Nevada.

The grass . . . cured rapidly in the hot sun and dry air. In the afternoon it was tied up in small bundles and piled on the wagon. There was a large load of it. We spent the day in making everything ready for the start toward the desert the next morning. The rumors that came back from there were very distressing—animals dying without number, and people suffering from prolonged thirst. (Margaret Frink, August 13, 1850)

In wet years the last tendrils of the Humboldt crept past Big Meadows to the Humboldt Sink to collect in a vast, shallow lake. In dry years the river didn't make it past Big Meadows, leaving the sink as an arid mudflat. Wet or dry, the Humboldt Sink was—and is—a wretched place: a vast plain of slurping mud when wet, or windswept dust and glaring salt flats when dry. James Pritchard described it as "a vast Quagmire or Marsh of Stagnant Saline & Alkali water mixed," and emitting "a most offensive and nauseateing effluvia." For Reuben Cole Shaw, it was "a mud lake ten miles long and four or five miles wide, a veritable sea of slime, a 'slough of despond,' an ocean of ooze, a bottomless bed of alkali poison, which emitted a nauseous odor and presented the appearance of utter desolation." For Margaret Frink, it was simply "the end of the most miserable river on the face of the earth."

INTERSTATE 80 follows the Humboldt River along its valley today. At 70 miles per hour, the mountain ranges shift in and out of view quickly. Now it takes all of 15 minutes to cover the distance that emigrants made each day. The ranges stack up to the horizon in rows, like teeth in a shark's mouth, their highest peaks tooth-white even in summer. Aridity takes over as your eye descends the sere slopes to the parched valley floor. There, in the middle of the valley, lies a twisting green ribbon—the Humboldt River. Pull off the highway and stop on one of the bridges that cross the river, and you gaze onto a world of busy, exuberant life. Willows, rabbit brush, berry bushes, and head-high grass crowd the channels. Coyote tracks crisscross the bottomlands. Rabbits peer from the thickets, insects flit, and birds chirp and dance about. A soft murmur from all of this activity rises up and mixes with the breeze. Walk away from the river 20 yards and dry silence closes in. The river is a thin, wet string of life humming gently in a silent realm of bare-rock mountains—the realm of the Great Basin.

But this appreciation of the river is made possible by modernity. Bridges today cross above the muck and mire; air-conditioning and windows shut out the relentless summer heat and mosquitoes; clean water and supplies (as well as slot machines, blackjack, and prostitutes—it's Nevada, remember) are rarely more than a few miles down the highway.

highest shoreline
of Lake Lahontan

Humboldt Sink

11.8 The Humboldt Sink—the end of the line for "the most miserable river on the face of the earth." The view looks east across the sink toward the West Humboldt Range. The sink is usually dry today because irrigation sucks up the Humboldt River before it gets here. During times of high rainfall or snowmelt, a vast, shallow lake may cover the sink for a few weeks or months at a time. Notice on the lower flanks of the West Humboldt Range the shoreline marks of prehistoric Lake Lahontan—a topic for the next chapter.

How different to be stuck with the Humboldt. The emigrants walked for three weeks and 350 miles along the wilderness river. They walked in summer's heat, utterly dependent upon the river yet utterly sick of it. Surely, the road west could offer up nothing worse than the Humboldt River.

Yet it would.

West Humboldt Range

THE WORST DESERT
YOU EVER SAW

The desert! You must see it and feel it in an August day, when legions have crossed it before you, to realize it in all its horrors. But heaven save you from the experience.

ELEAZER STILLMAN INGALLS, August 5, 1850

In the last 200 miles of the California Trail, some of the hottest, driest desert in North America runs up against one of the continent's largest mountain ranges—the Sierra Nevada. The mountain acts like a dam into the atmosphere, wringing moisture out of Pacific winds and leaving just a few drips for the Great Basin beyond. Donner Pass, on the crest of the range, catches an average of 54 inches of precipitation annually—more than Seattle—while the valleys of the Great Basin just a few miles east get about five or six inches. Between the Humboldt Sink and the nearest rivers flowing east off the Sierra Nevada—the Truckee and Carson rivers—lie 40 miles of desert. You couldn't lose your way on this Forty-Mile Desert in 1849. The route was clearly marked by dead and dying animals, and by abandoned wagons and their disgorged contents. Like the battlefields of great wars, the Forty-Mile Desert bore the scars of waste and death for years after the legions had gone, as Mark Twain saw in 1861:

> From one extremity of this desert to the other, the road was white with the bones of oxen and horses. It would hardly be an exaggeration to say that we could have walked the forty miles and set our feet on a bone at every step! The desert was one prodigious graveyard. And the log chains, wagon tires, and rotting wrecks of vehicles were almost as thick as the bones. . . . Do not these relics suggest something of the idea of the fearful suffering and privation the early emigrants to California endured?[1]

1. Twain traveled west by stagecoach in 1861 and settled for a time in the silver-mining town of Virginia City in present western Nevada. He tells the tales of his Western adventures in his 1872 book *Roughing It*.

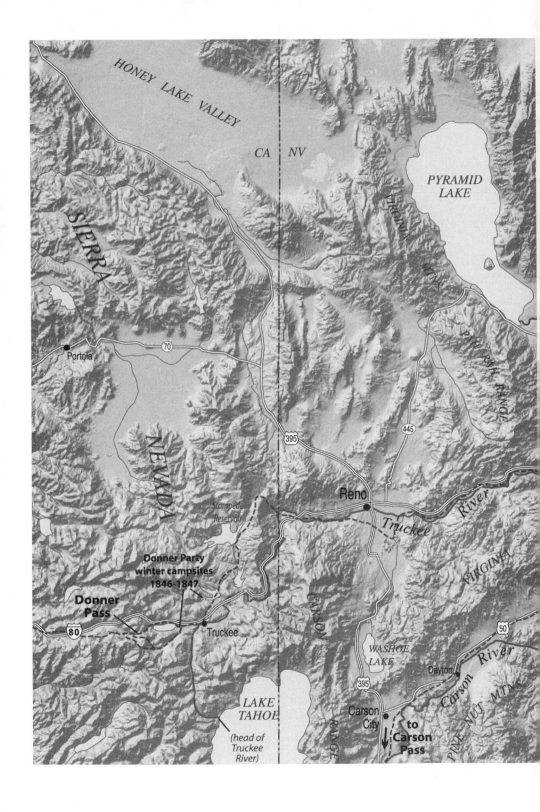

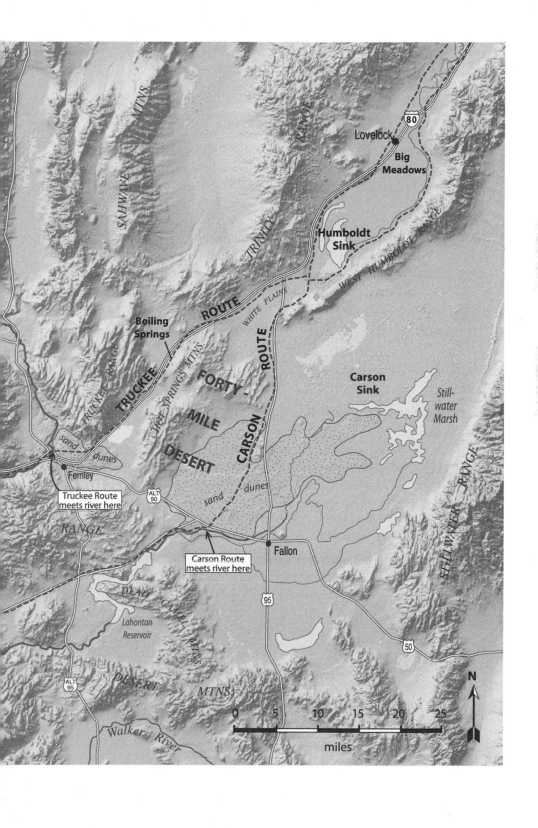

The problem with the Forty-Mile Desert wasn't that it was so far across. The dry run of the Sublette Cutoff (chapter 8) had been longer. The problem wasn't that it was so dry either. There were a few sources of water along the way, albeit boiling hot or highly saline. The problem was timing. Forced to squeeze their journey into the gap between the spring thaw of the Great Plains and the autumn snows of the Sierra Nevada, the emigrants had to cross the Great Basin in August and September—the hottest time of year. They had just come off the 350-mile Humboldt run, where they had endured three weeks of heat, dust, foul water, and Indian depredations. As 1850 emigrant J. S. Shepherd explained, the desert crossing "would not be as bad as it is, did it not follow close after the Humboldt River, the water of which has such a weakening influence on the stock." To make things worse, many emigrants had run nearly or completely through their food stores by this point. Others had lost their livestock and were packing on foot. Ragtag, gaunt, and bone-weary, they collected at the Humboldt Sink to prepare as best they could for the desert crossing.

> All are preparing and dreading to cross by far the worst desert we have yet met. They, perhaps, would not mind it, and neither would I, if we had plenty to eat; but here are hundreds already lamenting their antici-pated death, and suffering on the burning plain. Expect to find the worst desert you ever saw and then find it worst than you expected. (John Wood, 1850)

Most emigrants harvested grass at Big Meadows, a few miles above the sink, and then went on to the sink itself to tank up on water. In a wet year, such as 1850, there might be standing water on the sink—water "strong with salt and alkali and the color and taste of dirty soap-suds," Margaret Frink remembered. In drier years the sink was an arid mudflat. Then the only water came from stagnant pools or shallow wells dug along the edges of the sink. Augustus Burbank described the scene at these dug wells in August 1849:

> These wells although the water is not good, are like unto a fountain in a desert land. . . . This place is perfectly thronged all the time, & a con-tinual dipping & watering of animals is kept up, some departing for the desert & others arriving, quite a pell-mell, haw, gea & gea, haw. Quite a

12.1 (*previous pages*) From the Humboldt Sink, the emigrants took one of two routes across the Forty-Mile Desert—the Truckee Route or the Carson Route. It was about 40 miles to the Truckee or Carson rivers either way. More suffering and animal death occurred in these 40 miles than on any other part of the journey, mostly in the sand dunes that block the final miles to either river. Once the emigrants reached the rivers, they followed them upstream toward their final barrier—the high passes across the Si-erra Nevada.

12.2 Saline, fetid, algae-plugged pools like this one along the edges of the Humboldt Sink represent the final dregs of the Humboldt River—and the last water before the Forty-Mile Desert.

stench arises from the numerous dead cattle and horses that lay strewed around this vicinity & in the pool or slough, which is very offensive.

The dead cattle and horses did not go completely to waste. "Dead horses and oxen, in great numbers, with steaks cut out of their flesh, lay scattered over the land," John T. Clapp observed in July 1850, "and men, without a morsel to eat, were begging from wagon to wagon, offering all they had for a little dry bread." At 3,890 feet, the Humboldt Sink was the lowest elevation the emigrants had been since they passed Scotts Bluff on the Great Plains—1,300 trail miles back, and a world away.

SETTING OUT from the Humboldt Sink with every water cask brimming, everyone had to decide on one of two options for crossing the Forty-Mile

2. The Truckee River drains out of Lake Tahoe and flows down Truckee Canyon to Pyramid Lake—one of the few permanent lakes in the Great Basin. The river takes its name from Chief Truckee, an Indian who showed a group of 1844 emigrants the route across the Forty-Mile Desert to the river. That emigrant group—the Stephens-Townsend-Murphy party—became the first to take wagons across the Sierra Nevada, over what is now called Donner Pass. The Carson River heads near Carson Pass and drains to the Carson Sink. The river takes its name from mountain guide Kit Carson, who traveled along the river with John C. Frémont's exploring expedition in 1844.

Desert: the Truckee Route (blazed in 1844) or the Carson Route (1848). It was about 40 miles either way to the Truckee River or Carson River[2] (see fig. 12.1). Whichever route they chose, most set out in the afternoon and drove on through the night, hoping to get as far as possible before the heat of the next day. Either way, they stepped off into a kiln-dry world—one that took Great Basin bleakness to a whole new level. "Every thing around is sufficiently cheerless and desolate to depress the most buoyant temperament," Edwin Bryant wrote of the Truckee Route. "The sable and utterly sterile mountains, the barren and arid plain, incapable of sustaining either insect or animal, present a dreariness of scenery that would be almost overpowering in its influences, but for the hope of more pleasing scenes beyond."

Today the Truckee Route across the Forty-Mile Desert whisks by at highway speeds, for Interstate 80 rarely strays more than a mile from the emigrant trail. The landscape isn't much changed since 1849. Mummified mountains punch up through skirts of brown alluvium. Mud-cracked plains called playas (dry lakes) stretch utterly flat for miles between the ranges. Toe-deep with water after a rare heavy rainstorm, the playas lie dry most of the time, white with encrusted salt. On the Forty-Mile Desert, even the indomitable sagebrush gives way to tougher plants—greasewood mostly, along with shadscale, hopsage, and winterfat. Growing with tectonic slowness, the plants have barely begun to overtake the wagon ruts of '49, which still slash faint scars across the alluvial plains.

About 10 miles west of the Humboldt Sink, Truckee Route emigrants passed a bizarre cluster of rocks, looking like grotesque Christmas trees sticking up from the desert. "The rocks had a peculiar formation, such as I had never saw before or since," Heinrich Lienhard wrote of these tufa mounds. Tufa is a type of porous, friable limestone that forms in lakes. Groundwater rich in dissolved calcium seeps up through springs on lake floors. Where it hits lake water rich in dissolved carbonate, a happy chemistry ensues that precipitates knobby mounds of tufa. Stand by a tufa mound, in other words, and you stand on the floor of an ancient lake. In this case, the lake in question is Lake Lahontan—a Lake Ontario–sized body of water that once stood 500 feet deep over the Forty-Mile Desert.

A budding geologist quickly learns about the mind-bending environmental schisms that arise from the depths of time. I have taken students to Devonian reefs in Ohio and talked with them about the oceans that once covered the Midwest. It's intensely fun to see the eye-popping wonder of those who open the pages of their first outcrops and read the stories inside. Oceans in Ohio! Tropical forests in Antarctica! Himalayan-sized mountains stretching from Newfoundland to Atlanta! No novelist, no screenwriter—no lunatic even—could invent tales more fantastic than

12.3 Emigrants on the Truckee Route passed by these tufa mounds, visible today from Interstate 80 south of milepost 75. The large mound on the left stands 10 feet tall. These mounds, and others nearby, formed beneath Pleistocene Lake Lahontan. Tufa mounds are limestone (calcium carbonate) deposits that form on lake bottoms where groundwater seeps out through underwater springs. Algal growth apparently promotes tufa formation. During photosynthesis, algae pull carbon dioxide out of lake water, lowering its acidity and tipping the chemical balance toward precipitation of tufa.

those that have actually happened on this planet. But you get used to it, eventually. The wonder fades. Yet there's one geo-environmental mindbender that still knocks me flat—and that's the notion of 500 feet of fresh water over the Forty-Mile Desert.

The summer air aches with dryness in the Forty-Mile Desert today. Lay bread out for a sandwich and it desiccates halfway to toast before you can take a bite. The cloudless sky, flat and hard as an anvil, yields not a drop of moisture for months on end. But if you could bring the cooler, wetter climate that prevails in, say, southern Alberta down to the Great Basin, things would change. The current imbalance between precipitation and evaporation would shift. Today's anemic, saline streams would swell to tumbling rivers, filling the Great Basin with sparkling clear-water lakes.

This happened several times in the Great Basin during the past 2 million years. During this interval—known informally (and imprecisely) as the Ice Age—continental ice sheets repeatedly slithered south from the Arctic to cover huge swaths of North America and northern Eurasia. The ice melted back, and advanced again, perhaps 20 times or more. (Each advance of an ice sheet erases much of the evidence of previous advances, so it's hard to know exactly how many times the ice came and went.) The

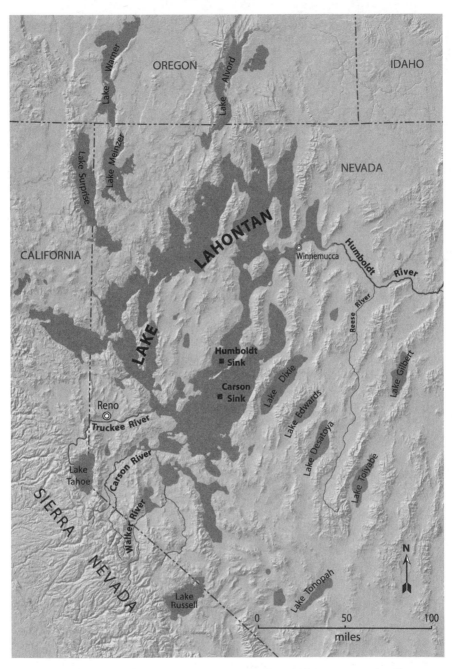

12.4 Lake Lahontan at its greatest extent 13,000 years ago, when wetter, cooler climates prevailed throughout the Great Basin. The lake stood nearly 500 feet deep over the Humboldt and Carson sinks and the Forty-Mile Desert. Lake Bonneville—more than two times larger than Lahontan—covered much of the eastern Great Basin at nearly the same time, and dozens of smaller lakes formed in other intermountain valleys as well.

ice scraped south as far as Idaho, Kansas, Ohio, and Long Island and scooped out holes that became great lakes, including the Great Lakes. The ice never advanced as far south as the Great Basin, but the glacial periods brought cooler and wetter climates to the region. Mountain glaciers, sustained by increased snowfall and cooler temperatures, began to grow on the highest Great Basin peaks and along the crest of the Sierra Nevada. The glaciers supplied abundant meltwater to Great Basin rivers, and the intermountain valleys of the Great Basin filled with lakes.

Today the shorelines of these Ice Age lakes stand as clear, horizontal bands, like bathtub rings, on the flanks of the mountain ranges. On the West Humboldt Range east of the Humboldt Sink, the highest shoreline of Lake Lahontan lies nearly 500 feet above the mud-cracked surface of the sink (see fig. 11.8). Like a chalk tracing around a body at a murder scene, the ancient shorelines show where the now-vanished lake once lay. At its maximum extent about 13,000 years ago, Lake Lahontan covered 8,600 square miles—larger than Lake Ontario—with a crazy, looping shoreline that, stretched out straight and including all the islands, would have reached from San Francisco to Nova Scotia. Lahontan had an even larger cousin, Lake Bonneville, which covered much of northwestern Utah at about the same time. Bonneville grew to some 20,000 square miles—a Lake Michigan—and at its highest level about 15,000 years ago spilled north out of the Great Basin into the Snake River in a series of monumental prehistoric floods (the Bonneville Floods) that rolled car-sized boulders as easily as a mountain stream moves grains of sand. Today's Great Salt Lake is the shrunken, saline remnant of Lake Bonneville. Looking east from downtown Salt Lake City to the Wasatch Range, you can spot Bonneville's highest shorelines 1,000 feet above street level.

With the latest retreat of the continental ice sheets about 12,000 years ago, dryness again descended on the Great Basin. The lakes shrank and mostly disappeared, leaving thick coatings of mud behind. The hooves and wagon wheels of the great westward migration kicked this old lakebed mud into dust. "One of the most disagreeable things in traveling through this country is the smothering clouds of dust," Alonzo Delano complained along the Humboldt River—whose valley 13,000 years ago lay mostly submerged beneath Lake Lahontan (as shown in fig. 12.4). "The soil is parched by the sun, and the earth is reduced to an impalpable powder by the long trains of wagons," he added. "We literally had to eat, drink and breathe it."

ABOUT 20 miles from the Humboldt Sink, halfway to the Truckee River, the emigrants pulled up at the Boiling Springs, an eerie collection of hot pools and roaring steam vents "where the ground shoots steam like the

12.5 One of dozens of steam vents that hiss and puff at the famous Boiling Springs along the Truckee Route. The vents line up northeast-southwest along several parallel faults that let scalding groundwater up to the surface. In many places here, the ground is too hot to touch with your hand. The smell of onions now joins the sulfurous stink of the vents, for a massive geothermal plant nearby uses the steam heat to dehydrate tons of onions for storage. The area lies just south of exit 65 off Interstate 80.

mouth of hell." Throughout the Great Basin, the Earth's thinly stretched crust lets hot mantle rock rise up close to the surface, so that hot pools and steam vents burst forth in dozens of places. The scalding water, although alkaline and reeking of sulfur, was potable when cooled. Consequently, as Franklin Starr saw on August 23, 1849, "there are a great many wagons lying by at the springs trying to cool water for the cattle." Few lingered for long. They watered the stock, packed up, and rolled on. For those who had left the sink the previous afternoon and traveled all night, the rising sun now began to tap on weary shoulders. Worse, some seven miles east of the Truckee River, boots, hooves, and wheels began to bog down in the sands of the Truckee Dunes.

Whichever route you took across the Forty-Mile Desert (Truckee or Carson), you faced a horrific pull at the end through sand dunes. The sand is there because of the life-saving rivers just beyond. Westerly winds pick up sand from the river bottomlands and spread it east across the desert—right in the path of anyone approaching from that direction.

The suffering that unfolded in the Truckee Dunes scorched every memory. Lucius Fairchild described the trail "lined with dead cattle, horses & mules with piles of provisions burned & whole wagons left for want of cattle to pull them through." Short on water and desperate to reach the river, most emigrants drove on through the heat of day. "The sun had risen and was high on the heaven when we got on the sand," William Woodhams remembered. "The dust flew in clouds, and we plodded wearily on, stopping every few minutes to bestow half a dozen blows on our poor tired animals that lagged the worst, some of them reeling as they went." Some unhitched their failing teams and headed for the river, hoping to return later for the wagons. Countless animals sank down in the sand and never got up. Exhausted and whipped beyond caring, they lay with glazed and puzzled eyes, "waiting for death," John Edwin Banks observed, with "no possibility of other relief." The stench of decomposing flesh rose in appalling clouds. California-bound emigrants were no strangers to animal suffering and death—but never on a scale like this. "All our traveling experience furnishes no parallel," Bennett Clark admitted—and this with the Humboldt River experience still fresh in mind.

Interstate 80 swings well south of the emigrant trail through the Truckee Dunes today. Only the hiss of windblown sand disturbs the graveyard quiet now. Jeep tracks follow the trail across the dunes toward the river. The remnants of the grim march, the bones and abandoned wreckage, are gone—collected, decayed, or buried by the shifting sand. Modern artifacts are abundant enough, though, especially as you approach the Truckee River: beer bottles, moldering mattresses, rusted appliances—the detritus of the nearby towns of Wadsworth and Fernley. Shotgun shells (the droppings of the Lesser Redneck) abound as well. The Truckee Dunes, fossilized, will present future archaeologists with an intriguing stratigraphy: bleached oxen bones and rusted wagon parts overlain by bullet-holed appliances and faded porno magazines—the stratigraphy of Manifest Destiny.

Approaching Truckee River, the trail through the dunes converges again with the interstate and its billboards—including one for *Pioneer Crossing Casino—Blazing New Trails to Fun!* Then, cresting the last sandy ridge, a bright green band comes into view—the cottonwood trees along the Truckee River. "If ever I saw heaven, I saw it then," Lydia Waters wrote in 1855. Skidding down the final dunes, the emigrants came at last to the river. "It is wonderful to see cattle rush into the water," John Edwin Banks remembered. "O how delicious! I know it was to me."

The Truckee River runs as cold, clear, and lovely now as it did in emigrant days, 50 feet wide and knee-deep, fed by everlasting Sierra Nevada snows. Towering cottonwoods line its banks, catching the wind with a sigh and casting a corridor of cool shade through summer's heat. These were the first tall, shady trees most emigrants had seen since the Green River some 800 miles back. "No one can imagine how delightful the sight

12.6 Possibly the most beautiful sight—for an emigrant at least—along the entire 2,000-mile California Trail was this first glimpse of the cottonwoods along the Truckee River. The trees come into view as you crest the final sand dunes along the Truckee Route across the Forty-Mile Desert. The dunes formed a ghastly killing field in emigrant days, paved with dying, dead, and putrefying animals and abandoned possessions. The white rectangles in the far left distance are billboards along Interstate 80, which follows the emigrant trail and the Truckee River upstream through Truckee Canyon toward Donner Pass.

of a tree is after such long stretches of desert," Elisha Perkins explained, "& what a luxury . . . to lay down in their shade & make up our two nights loss of sleep, & hear the wind rustling their leaves & whistling among their branches." Today these cottonwoods beautify Wadsworth, an otherwise-dismal hamlet of boarded-up houses and run-down cars washed up against a bend in the river.

IF YOU DIDN'T take the Truckee Route across the Forty-Mile Desert, you took the more southerly Carson Route. Blazed in 1848, the Carson Route quickly overtook the Truckee as the most popular trail for gold rush emigrants. It had the advantage of an easier ascent through the eastern Sierra Nevada foothills (in contrast to the route up Truckee River Canyon, which was horribly rough and involved two dozen crossings of the river). The Carson Route offered no advantage in crossing the Forty-Mile Desert, though—in fact, it may have been worse. A few dug wells supplied some water ("intensely brackish, bitter with salt and sulphur")

12.7 The Truckee River—goal and salvation of emigrants crossing the Forty-Mile Desert. The view is west (upstream) close to where emigrants first came to the river. The town of Wadsworth lies just out of view to the right.

a few miles beyond the Humboldt Sink.[3] After that it was just a long, dry push to the river, much of it through deep sand. The sand dunes on the Carson Route stretched farther than those of the Truckee Route, and consequently the animal death and wreckage seem to have been worse. Eleazer Stillman Ingalls described what it was like in the Carson Dunes on August 5, 1850:

> The sand hills are reached; then comes a scene of confusion and dismay. Animal after animal drops down. Wagon after wagon is stopped, the strongest animals are taken out of the harness; the most important effects are taken out of the wagon and placed on their backs and all hurry away, leaving the wagons, property and animals that, too weak to travel, lie and broil in the sun.

3. You can visit some of these dug wells along the Carson Route today. I drank from the so-called Double Wells (about five miles south of the Humboldt Sink) on a hot September day much like the emigrants would have experienced. The water tasted of salt and dish soap mixed with algae. It took a whole beer to take the flavor away—and then a second to toast the hapless souls who had to drink the stuff to survive—and then a third to toast the beauty of cold beer on a hot day in the desert.

John Hawkins Clark, crossing in 1852, probably witnessed the dunes at maximum carnage. Freshly dead and dying animals joined the moldering corpses of years past. Dead stock "filled the entire roadside; mostly oxen, with here and there a horse and once in a while a mule."

> In many places the teams lay as they had fallen; poor beasts—they had struggled on over the mountains, plains, and through the sands of the barren desert for days and weeks with but little or no food, but still with strength sufficient to make this their last effort to gain a haven of rest. Good water and plenty of food lies just beyond; but alas, strength failed and here they lie, sad memorials of a grand crusade to "the land of gold."

Beyond the animal death and waste, the sheer volume of abandoned goods astonished everyone. The Carson Dunes looked as if a routed army had passed through. "The destruction of property upon this part of the road, is beyond all computation," Franklin Langworthy declared.

> The desert from side to side is strewn with goods of every name . . . log chains, wagons, wagon irons, iron bound water casks, cooking imple-ments, all kinds of dishes and hollow ware, cooking stoves and utensils, boots and shoes, clothing of all kinds, even life preservers, trunks and boxes, tin bakers, books, guns, pistols, gunlocks, gun barrels. Edged tools, planes, augers, and chisels, mill and cross cut saws, good geese feathers in heaps, or blowing over the desert, feather beds, canvas tents and wagon covers.

The Carson Route across the Forty-Mile Desert easily trumped the miseries of the Humboldt River. "For many weeks [along the Humboldt] we had been accustomed to see property abandoned and animals dead or dying," Margaret Frink explained. "But those scenes were here doubled and trebled. As we advanced, the scene became more dreadful. The heat of the day increased and the road became heavy with deep sand. . . . The stench arising was continuous and terrible."

There was nothing to do but press on through the acres of putrefying meat. "The day was oppressively hot, and the burning sands reflected the rays of the sun to such a degree that it appeared like suffocation at once." William Kelly saw men delirious with thirst, "howling for water," and throwing themselves down "in a fainting state under the shade of the wagons." In the gold rush years, traders from California set up camp along the Carson River, and some of them carried water out to arriving emigrants. They charged a dollar or more per gallon (roughly $20 today), and their customers rarely complained. "After the nauseous stuff of the Humboldt 'sink,' this spring water was more than an ordinary luxury" for Margaret Frink.

Finally, a hint of green—the cottonwoods along the Carson River—appeared through the waves of heat. Jasper Hixson watched men "rush up, half crazed with thirst and hunger and embrace these noble old trees and weep as children, and bless God for their deliverance." "Before we got to the river," James Carpenter remembered, "we were so nearly famished that our tongues were Black and neither could talk. We got to the river and such water we could not drink our tongues were so swollen, but we burried our faces in the clear bright water guzeled it us [up] as best we could then waited a few minutes and guzel again."

Arrival at the Carson River staved off the threat of starvation that had loomed for many along the Humboldt River. Traders from California during the gold rush years stood ready to greet emigrants stumbling out of the desert and relieve them of their burden of dollars. Thomas Christy fumed about the outrageous prices, "all to strap the poor starving emigrant." Luckily for those with no money, not every merchant along the Carson was out to bilk the needy. "At Ragtown [a shantytown that sprung up where the trail met the river], to our great surprise, we found an abundant supply of flour from California," Franklin Langworthy wrote. "The flour was sent here by the Benevolent Society of Sacramento City. The agent, who had a large cloth tent, sells the flour for twenty-five cents per pound to those who have money, and gives twenty pounds to each who is destitute of cash."

Safe on the Carson, many emigrants paused to rest for a day or more, luxuriating in the shade of the cottonwoods and drinking the delicious river. "Its water was clear, cool, and pure, free from salt or alkali, as different from the Humboldt soap-suds [as day?] is from night," Margaret Frink wrote. The next task for many (after their animals had recuperated) was to drive back across the sand and retrieve abandoned wagons and goods. "God of Heaven! Could human suffering appease thy wrath, the world would soon be forgiven," John Wood wrote of the desert crossing. For Thomas Christy, the Forty-Mile Desert was simply "the awfulest country that a man ought to travel."

BLAME THE Sierra Nevada for the Forty-Mile Desert. Its rain shadow casts aridity across the Great Basin, and the dryness intensifies the closer you get to the mountain. Ahead to the west, this final barrier loomed into view as the emigrants plodded west up the valleys of the Truckee or Carson rivers. William Kelly described the emerging sight of the mountain from the Carson Route:

> From the summit of the rise we got the first good and distinct view of the
> Great Sierra Nevada range, stretching beyond the scope of vision, north
> and south, with pointed snow-capped peaks between us and the land of

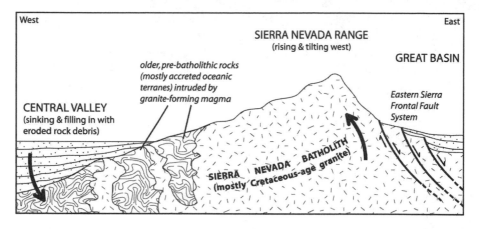

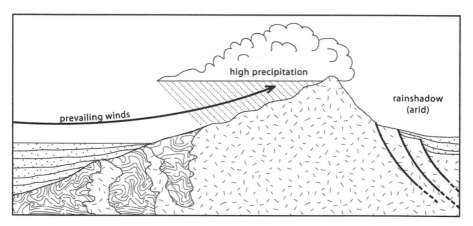

12.8 The Sierra Nevada and Central Valley together comprise a large, rotating block of rock tilted toward the west. The uptilted eastern side of the block forms the range, while the down-tilted western side forms the Central Valley, gradually filling with rock debris eroded from the range. The range rises along the Eastern Sierra Frontal Fault System, a band of active earthquake faults along its east side. The Sierra Nevada's rapid rise and westward tilt give it a distinct asymmetry, with a gentle western slope and a steep eastern escarpment. The range traps moisture from prevailing westerly winds, allowing little precipitation into the Great Basin to the east.

promise, reminding me of days gone by, the garden walls capped with glass to prevent naughty boys from stealing the rich fruit beyond them.

The Sierra Nevada range runs 400 miles long and 60 miles wide, with a glacier-gnawed ridgeline that boasts eleven peaks higher than 14,000 feet. The range and the adjacent Central Valley together comprise a long, rectilinear block of crust, aligned nearly north-south and tilted to the west. The up-tilted part of the block forms the mountain; the down-tilted part forms the Central Valley. The mountain's asymmetry reveals this westward tilt. The west slope forms a long, gentle ramp, while the east slope presents a steep, rugged escarpment. The crest of the range—the line of high peaks and ridges that divide west-flowing, Pacific-bound streams from those that flow east into the Great Basin—stands not in the middle of the range but next to the eastern escarpment. For instance, Mount Whitney (the highest point in the lower 48 states at 14,495 feet) lies only 10 miles from the Owens Valley to the east but 50 miles from the Central Valley to the west.

The Sierra Nevada is a mere babe among world mountains—young even by the youthful standards of the North American Cordillera. It began to rise not much more than 5 million years ago—a moment as recent in Earth history as three months ago is to the American Revolution. The evidence for its youth comes from fossil streambeds on the west side of the mountain. These old streambeds slope down to the west more steeply than do today's streams, having apparently gained extra slope as the range rose and tilted west. (The amount of extra slope translates to 6,000 to 8,000 feet of uplift along the eastern escarpment.) About 10 million years ago, some of these fossil streambeds became channels for lava that poured out of volcanic centers that are now *east* of the mountain. The lava, of course, could not have flowed uphill over the mountain. Therefore, there must have been no Sierra Nevada in the way when the lava flowed 10 million years ago. Step forward in time to 5 million years ago, and the sediments of the Central Valley record a rapid eightfold jump in sand and gravel accumulation rates, as if a convoy of dump trucks had suddenly arrived. The surge in sediment deposition signals the rise of a big sediment source to the east—something like a rising mountain. Conclusion: there was probably no Sierra Nevada 10 million years ago, but by 5 million years ago the mountain was rising and tilting west in earnest.[4]

And it isn't done yet. The Sierra Nevada still grows. Earthquakes are its growing pains. The quakes occur mostly along the Eastern Sierra Frontal Fault System—a 10-mile-wide band of faults that parallels the range's eastern escarpment. Whenever a fault along this zone shifts, it lets the range jerk upward a bit. Usually these spasms are small, lifting the range a fraction of an inch. Sometimes they are cataclysmic, pitching it several feet higher in a virtual instant—as happened on March 26, 1872. At 2:25 a.m. that morning, the Sierra Nevada ripped loose from its moor-

---

4. Or maybe not. The age of the Sierra Nevada is actually a topic of some controversy. While the "young mountain" theory I give here has held sway for years, some evidence now suggests that the Sierra Nevada may have existed as a major topographic feature as much as 50 million years ago, with an additional major growth spurt during the last 5 million years. See the Notes section for references.

ings along a fault near the small town of Lone Pine, in California's Owens Valley. Within seconds the mountain was six feet taller and 23 residents of Lone Pine lay dead, crushed in their sleep under collapsed roofs and walls. The earthquake jolted the naturalist John Muir out of bed in his Yosemite Valley cabin 110 miles away, where he listened to gigantic rock avalanches crashing down into the moonlit valley. As far away as San Diego, mechanical clocks shook to a stop.[5]

The scarp—the surface rupture of the fault that broke in the 1872 quake—rises like a raw scar on the west side of Lone Pine today. It forms a cliff up to 12 feet high, trending north-south, cleanly cutting the otherwise smooth surface of the alluvial fans that sweep down from the Sierra Nevada's eastern escarpment. The scarp is probably the work of several quakes, with 1872 being the latest. A gully filled with distinctive, angular rocks eroded from the nearby Alabama Hills heads east away from the foot of the scarp. To find the same gully on the other side, you need to scramble up the scarp *and* step 35 feet to the north. The gully once ran straight, before the scarp cut it in two. Its offset shows us that, with each earthquake, the Sierra Nevada grows not just *higher*, but leaps *north* as well. In the 1872 quake, the mountain rose by 6 feet and shifted north by 15 feet.

There's nothing strange about this. Practically all of the major faults in the western Basin and Range, when they shift with the stretching crust, let the rock on their west sides move north or northwest as well as either up or down. The faults tell us that, with each earthquake, the western Basin and Range pulls northwest away from the rest of North America. The reason—as we saw in chapter 10 (you might want to look again at fig. 10.9 here)—is that the western United States is tied by friction to the northwest-moving Pacific Plate. That plate, creeping northwest about two inches per year, is dragging several hundred miles of the western United States along with it. The Sierra Nevada is the biggest slice of rock caught up in this shuffle.

Luckily for anyone living within 50 miles of the Sierra Nevada's eastern escarpment, the mountain usually grows in milder spasms than it did on March 26, 1872. Still, every year it grows a bit more. Small quakes, most detectable only with seismographs, pop off weekly along the East-

5. This earthquake—the 1872 Owens Valley quake—happened before modern seismographs, so we can't determine its exact magnitude. Indirect evidence (the amount of offset along the fault, the size and length of the scarp, and personal accounts of shaking and damage) suggests that it was one of the three largest earthquakes to have hit California in historic times. The other two were the 1857 Fort Tejon quake and the 1906 San Francisco quake. All of these quakes probably had magnitudes between 7.8 and 8.2 on the Richter scale.

12.9 The east face of the Carson Range, uplifted along the Genoa Fault, typifies the Sierra Nevada's steep eastern escarpment. The lower photograph looks north along an exposed portion of the fault in a quarry at the base of the range near Genoa, Nevada. Grinding movements along the fault have masticated the once-solid granite into loose chunks and rock powder (*inset*). The Carson River flows close by the escarpment here because, with each earthquake, the Carson Valley drops and tilts toward the escarpment even as the mountain rises. The rapid and geologically recent rise of the Sierra Nevada along faults like the Genoa Fault made it the most daunting topographic barrier along the California Trail.

ern Sierra Frontal Fault System. Each one records a little shift that lets the mountain up a bit and/or north a bit. No single fault raised the Sierra Nevada. The mountain stands high because of the cumulative—and ongoing—work of an army of faults.

The best place I know to see, and touch, one of the faults that raised the Sierra Nevada is at the base of the Carson Range in western Nevada. The Carson Range is part of the Sierra Nevada's eastern escarpment. It rises like a wall 4,000 feet from the floor of the Carson Valley just to the east. At its foot lies the Genoa Fault—the one responsible for enlofting this particular segment of the Sierra Nevada. The fault plane—the surface along which the mountain jolted upward—lies exposed in a small quarry at the base of the range, one mile south of the town of Genoa. Its sheer face is scarred with slickensides—parallel grooves scraped on the rock as the range rose. Run your fingers over these gouges and you feel the pulse of a living mountain. The rock *was* classic Sierra Nevada granite, the same salt-and-pepper stuff that makes up the backdrops of Yosemite Valley and Kings Canyon. But you would hardly recognize it here. Eight thousand feet of grinding motion along the fault—the cumulative slip of thousands of earthquakes—has pulverized the granite so thoroughly that you can dig into it with your finger. Solid granite has become granite flour, ground between geologic millstones. Only the pea- to plum-sized bits of surviving granite that crumble from the pulverized mess reveal the rock for what it was.

Sierra Nevada granite is well-traveled rock. Most of it began below where Las Vegas is now, before the crustal stretching of the Basin and Range Orogeny dragged it 150 miles west-northwest to its present location. It began as blobs of magma that rose from the subducting Farallon Plate, mostly between 120 and 80 million years ago. Some of the magma jetted out as lava and volcanic ash to build a towering arc of volcanic mountains that erosion has long since erased. But much of it gurgled to a stop about 5 to 10 miles underground. There it congealed into bulbous plutons of granite. The accumulating plutons eventually welded together into the 400-mile-long Sierra Nevada Batholith. Then, about 15 to 20 million years ago, the Basin and Range Orogeny kicked into high gear and the batholith—still deep underground—left the Las Vegas area and headed west-northwest with the stretching crust. About 5 million years ago, the Genoa Fault and its countless cousins along the Eastern Sierra Frontal Fault System began to let the mountain up, exposing its batholithic core to view.

AS A DIRECT consequence of the Sierra Nevada's recent uplift, the emigrants creaked toward a mountain escarpment that dwarfed anything in

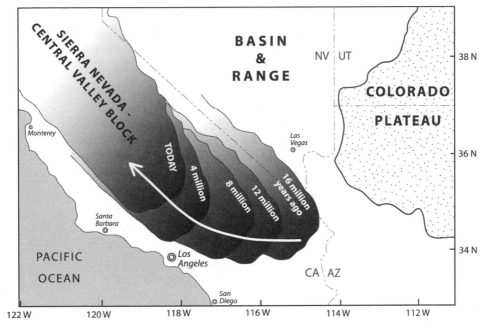

12.10 The well-traveled Sierra Nevada. Restoration of the Sierra Nevada–Central Valley crustal block to its original position (based on adding up movements on Basin and Range faults between the Sierra Nevada and the Colorado Plateau) shows that it originally lay near Las Vegas. As the crust stretched during the Basin and Range Orogeny, the block came along for the ride, traveling more than 150 miles west-northwest. This stretching continues today—as figure 10.9 shows.

their experience. With each step west, rock rose higher until it seemed to fill the sky. The Sierra Nevada came up too fast to let preexisting rivers cut antecedent paths through it; and it came up so recently that its fault-splintered eastern escarpment still rises sharp and sheer. There is no avoiding the eastern escarpment. To get to California (unless you care to add hundreds of miles to your journey by going around), you must make a frontal assault. Even today only one interstate (I-80) and a handful of paved roads cross the mountain east-west. And of course in emigrant days, there were no roads in the modern sense—only boulder-littered wagon trails. For Elisha Perkins, rolling up to the escarpment "was exactly like marching up to some immense wall built directly across our path." The view, though intimidating, was incomparably grand. The ragged teeth of the range, flossed by Ice Age glaciers and whitened by eternal snows, soared magnificent overhead. James Pritchard thought it "one of the grandest and most sublimely picturesque Sceneries that I ever beheld." Margaret Frink and her husband, Ledyard, were also impressed.

"We never tire of looking at the great mountains we are soon to climb over," she wrote. Dense forests of cool pines cascaded down the eastern slopes, beckoning the desert-weary travelers upward and onward. "Trees once more!" exulted Andrew Grayson.

> Oh! how the very sight of them cheered are [our] worn spirits. As we hurried to get among them, how gladly did we hail the change. As we entered the majestic woods, the breath of the forest was animating to us. What a feeling of freshness diffused itself into our whole being as we enjoyed the pleasures of the pathless woods. . . . What a change from the wild, wide wastes over which we had traveled for months.

INTO THE LAND
OF GOLD

I wish California had sunk into the ocean
before I had ever heard of it.

JAMES WILKINS, September 9, 1849

It was within sight of the Sierra Nevada's eastern escarpment—the final barrier to California—that 36 members of the Donner party died in the winter snows of 1846-47. The evidence confronted many later emigrants who took the Truckee Route toward Donner Pass.

> We arrived at the place where the Donner party perished, having lost their way and being snowed in. Most of them suffered and died from want of food. This was in 1846. Two log cabins, bones of human beings and animals, tops of the trees being cut off at the depth of snow, was all that was left to tell the tale of that ill-fated party, their sufferings and sorrow. (Sallie Hester, September 14, 1849)

The specter of the gaunt, shivering Donners sucking on the bones of their comrades haunted everyone who passed though the remains of the winter camps. All around, "stumps from ten to twenty feet high that were haggled off at the snow line" stood testimony to the snow depths that had sealed the party's fate. Over the winter of 1846-47, several rescue teams from California had reached the stranded emigrants, bringing meager rations and pulling out a few of the survivors. But the snowstorms roared through again and again, and despite heroic efforts, the last of 45 survivors did not reach safety until the end of April 1847. Trapped for more than five months, the marooned Donners ate up all of their cattle and dogs. They ate field mice. They gnawed on twigs and bark. They boiled ox hides to make glutinous soups—and then cut the ox hides into strips and ate those. Then, as some died, their bodies became meals for those still living. Wandering through the camps on August 21, 1849, Wakeman

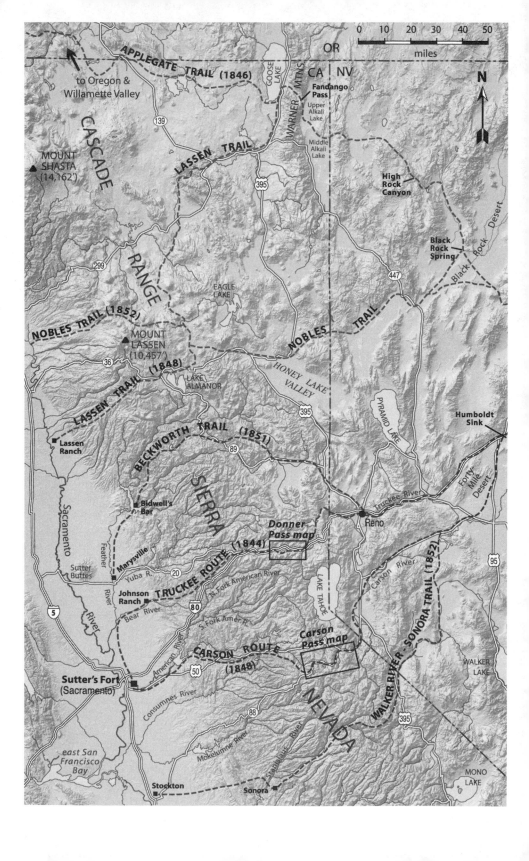

Bryarly claimed to have "found many human bones. The skulls had been sawed open for the purpose, no doubt, of getting out the brains, & the bones had all been sawed open & broken to obtain the last particle of nutriment." At Alder Creek, one of two Donner campsites,[1] the U.S. Forest Service has built a picnic area and dubbed it the Donner Camp Picnic Ground, oblivious, apparently, to the macabre allusion.

LEAVING THE dark woods around the "Cannibal Cabins," as some gold rushers called the dilapidated Donner shelters, the emigrants emerged onto the shores of Truckee Lake (now called Donner Lake). There they gazed west at the final great barrier to California—a wall of rock rising 1,160 feet to Donner Pass. "Standing at the bottom and looking upwards at the perpendicular, and in some cases, impending granite cliffs," Edwin Bryant remarked, "the observer, without any further knowledge on the subject, would doubt if man or beast ever made a good passage over them."

The final approaches to most passes over the Sierra Nevada run up the back walls of cirques—large, amphitheater-like bowls, some more than a mile across, scooped out at the heads of valleys by Ice Age glaciers. Like an ant trying to climb out of an ice-cream scoop, anyone climbing toward a Sierra Nevada pass hits the steepest terrain—the high back wall of the cirque—at the end, usually in the last 1,000 to 2,000 vertical feet. Donner Lake sits in the bottom of a cirque that rises west toward Donner Pass in a series of huge, irregular steps, like a giant's staircase. The thick pine forests around the lake thin quickly as you climb toward the pass. The remaining trees sink their roots into the cracked granite and hang on, braced against the cold rivers of wind that pour east over the pass. The wind snubs off the west-side branches and sculpts the trees into east-streaming weathervanes. Lightning regularly hacks off the treetops so that some are nothing more than living stumps. Donner Pass itself—a

1. The 81 members of the trapped party holed up in two winter camps, with about two-thirds of them setting up camp at the east end of Donner Lake (at the site of present Donner Memorial State Park), and the others staying at Alder Creek about five miles to the northeast (see *fig. 13.2*).

13.1 (*previous page*) Approaching the Sierra Nevada, the overland trail system diverged like strands unraveling at the end of a rope. The strands offered about six options for crossing the mountain, including roundabout routes around the northern end of the range (the Lassen and Nobles trails) or trails that went directly across the center. The oldest Sierra-crossing trails are the Truckee Route (blazed in 1844) and the Carson Route (1848). The Truckee Route over Donner Pass handled most of the pre–gold rush emigration. The Carson Route was the favored trail of the gold rush years, with the Lassen Trail also handling a lot of traffic. Note the locations of the Donner Pass and Carson Pass maps that appear on the next pages.

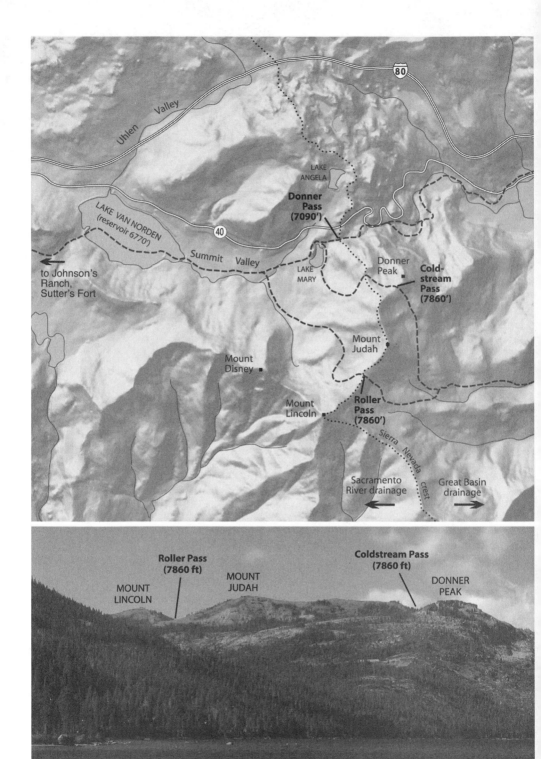

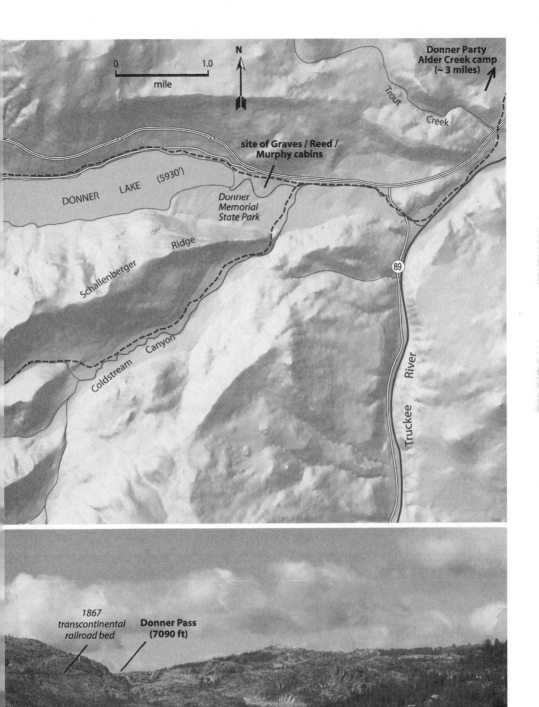

0   1.0

mile

N

Donner Party
Alder Creek camp
(~ 3 miles)

Trout
Creek

site of Graves / Reed /
Murphy cabins

DONNER   LAKE   (5930')

Donner
Memorial
State Park

Ridge

89

Schallenberger

Truckee
River

Coldstream   Canyon

1867
transcontinental
railroad bed

Donner Pass
(7090 ft)

DONNER LAKE (5930 ft)

quarter-mile-wide saddle between higher peaks—is mostly bare, fractured granite laced with angular inclusions of dark country rock (the native schist that was in the area when the granite-forming magmas barged in from below). Chunks of this schist crumbled from the walls of the underground magma chambers to be entombed in the crystallizing granite like pieces of fruit in congealing Jell-O. They are called xenoliths—literally "foreign rocks." It's an odd term considering that the granite is really the foreigner—the invader that overwhelmed the native rock and attempted to assimilate it.

Two stories of transcontinental passage are written at Donner Pass. In 1844 an emigrant party lead by Elijah Stephens took the first wagons west over the pass into California. By showing that wagons could go over the Sierra Nevada, the Stephens party opened the emigration floodgates to California. (Stephens Pass was later renamed Donner Pass.) Just over two decades later, the transcontinental railroad came through the pass in the other direction. From 1866 to 1867, Chinese laborers for the Central Pacific Railroad punched a one-third-mile-long tunnel through the granite below the pass. Using hand tools, black powder, and nitroglycerin, the Chinese dug the tunnel and blasted out notches in the sheer mountainside to make the railroad bed. They bridged one low section with a massive wall constructed of cut granite blocks. Today's railroad passes through a newer tunnel, but the original Chinese cut-block wall and tunnel stand at Donner Pass today as a testament to human tenacity—and the power of rice.

At 7,090 feet, Donner Pass is one of the lowest passes over the Sierra Nevada. Yet it was one of the hardest passes for the emigrants because of the high, steplike ledges that blocked their ascent. Sometimes the only way to get up and over the ledges was to take the wagons apart and haul them up piece by piece.

> We came to a rim rock ledge where there was no chance to drive up, so the wagons were taken to pieces and hoisted to the top of the rim rocks with ropes. The wagons were put together again, reloaded and the oxen which had been led through a narrow crevice in the rim rock were hitched up and went on. (Benjamin Bonney, 1846)

Another strategy was to cut long poles from sapling trees and lay these up across the ledges (visualize leaning poles against a flight of stairs). Then they would slide the wagons up the poles.

13.2 (*previous pages*) Donner Pass area map. By 1849 the Truckee Route had split into three options for crossing the dividing ridge of the Sierra Nevada: Donner Pass, Roller Pass, and Coldstream Pass. The photograph looks west to the three passes from the east end of Donner Lake near Donner Memorial State Park.

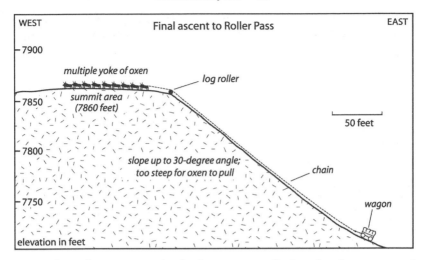

WEST   Final ascent to Roller Pass   EAST

7900

multiple yoke of oxen   log roller

summit area
(7860 feet)

7850

50 feet

7800

slope up to 30-degree angle;
too steep for oxen to pull   chain

7750   wagon

elevation in feet

13.3 The log-roller strategy used to haul wagons up Roller Pass, based on accounts of the 1846 Brown Party.

> When we came to the benches of rock six and eight feet straight up and down we would unyoke our oxen, drive them around to some low place, get them above the bench, yoke up the oxen. In the meantime some of us would cut some large poles strong enough to bear up our wagons and lay them up on the rocks. Then take enough chains to reach back to the wagons, hitch to the end of the tongue, and pull the wagon up. (David Hudson, 1845)

It was brutal work for man and beast, and as word spread about the terrible climb, emigrants began to look for alternatives to Donner Pass. Between 1846 and 1848, scouts reconnoitered two new routes over the mountain: Roller Pass and Coldstream Pass. The approach for both passes goes up Coldstream Canyon—the next glacial valley to the south of the Donner Lake valley. Although both passes are nearly 800 feet higher than Donner Pass, their final approaches—although steep—are not ledgy. If you had enough animal power, you could conquer these passes without emptying or dismantling your wagons. At Roller Pass, the final 400 feet form a 30-degree slope of jagged boulders, "as steep as the roof of a house," Joseph Hackney observed (and without exaggeration, for to look down that slope from Roller Pass today is to feel the same tipsy vertigo you get when standing on the apex of a steep roof). Oxen could not pull wagons directly up such a grade. Instead, the animals were un-hitched from the wagons and driven to the top, where they could heave along the flatter ground at the pass. Some emigrants placed log rollers at the top of the slope. They yoked the oxen to hundreds of feet of chain, and then passed the chain over the rollers and down the slope to a waiting wagon. Then, under shouts and cracking whips, as many as 12 yoke of oxen hove to on each wagon. The taut chain gnashed into the logs and, one by one, the wagons creaked upward to gain — at last — the roofline of the mountain.

The reward for a job well done was the spectacular view from California's glacier-carved rooftop. "The aspect of the country wild in the extreme," John Edwin Banks marveled. "Rugged naked rocks on every hand, occasionally relieved by a majestic pine which has seized a foothold. Here the grizzly bear may roam lord for ages."[2] The emigrants now stood on the divide between Pacific-bound and Great Basin–bound streams. They shed no tears over leaving the Great Basin. "We felt a real relief," John Steele wrote, "in bidding farewell to the mountains, valleys, and deserts of the great interior, with its adventure, romance, tragedy, sorrow, suffering and death — scenes which will linger in our minds as memorials of our journey across the plains." All eyes now looked west for the first glimpse of California — land of hopes and dreams. "As I stood there . . . and cast my eyes westward," A. J. McCall wrote from the mountain crest on September 7, 1849, "a picture of wonderful grandeur and magnificence was spread out before me."

> Below were a succession of innumerable pine-covered mountain peaks, growing less and less until they disappeared in a broad, yellow valley sweeping north and south until lost to view, and beyond another range of mountains. This was the far-famed Sacramento Valley, nearly a hundred miles distant. The purity of the atmosphere rendered vision almost illimitable, showing every line and shadow distinctly.

The clear view west showed all too well the work that still lay ahead. McCall continued:

> Such fearful gorges, such deep, deep ravines and canyons were fearful to behold. I was fairly appalled at the work before me, but that others had made the descent in safety I should have despaired. The climbing of this mountain was small work, as it seemed to me, compared with the descent.

Yet "it is will and pluck that overcomes mountains," he decided. And with that, he began the descent toward the land of gold.

SIXTY MILES AWAY, Carson Route emigrants plodded toward their own Sierra Nevada rendezvous. From the Forty-Mile Desert, the trail heads west, upstream along the Carson River. It then swings south with the

---

2. Banks could not know how quickly California's settlement would change things. The last grizzly in the state was shot dead in 1922.

river through the Carson Valley, along the 4,000-foot-high escarpment of the Carson Range, where the Genoa Fault (fig. 12.9) has done its part to raise the Sierra Nevada. South of the Carson Valley, the fun begins. The trail right-angles uphill into sheer-walled Carson Canyon, where the Carson River tumbles down a six-mile chain of foaming cascades. Boulders the size of bears and buses plug the canyon bottom. For George Read, the rock-riddled trail up Carson Canyon was "most decidedly the worst road I ever passed over with a team." Forty-niner William Kelly described "mules and wagons staggering over confused piles of rocks, where a goat could scarcely walk with confidence." Where they could not go around the boulders, the men pried and pulled the wagons up and over with ropes, levers, and crowbars. The impending granite cliffs on both sides threatened to send more rocks down onto their heads. "It made one's flesh creep," William Kelly wrote, "to look up and see huge crags suspended . . . wanting only the vibration of an echo to break the frail ligatures, and grind you into eternity." For William Johnston, the ascent of Carson Canyon was accomplished only through "the wildest hallooing, the loudest of whip cracking, and the most extraordinary profanity that ever saluted ears, whether of dumb beasts or of men."

After six hard-fought miles, they reached the top of Carson Canyon and emerged into Hope Valley—an open, level basin 7,100 feet above sea level. Here the route leveled out and everyone breathed easier for a time. Cool, green Hope Valley, spangled with wildflowers, could not have stood in starker contrast to the blasted desert landscape so recently passed. But they were still on the wrong side of the mountain, and two more tough climbs lay ahead—8,570-foot Carson Pass and 9,500-foot West Pass.

Like the routes up to most Sierra Nevada passes, the ascent to Carson Pass[3] goes up the back wall of a glacier-carved cirque. From Hope Valley, the trail climbs steadily to a small lake at the base of the cirque. From there, the cirque wall rises 720 feet in scarcely one-third of a mile to the

3. Carson Pass and the Carson Route take their name from scout Kit Carson, who passed through the area in 1844 with John C. Frémont's exploring and mapping expedition. But the Frémont expedition didn't bring wagons through the mountains here. For the blazing of a wagon route, the emigrants had the 1848 expedition of the Mormon Battalion to thank. In 1846, at the request of President James K. Polk, Mormon leader Brigham Young sent a group of young Mormons from Salt Lake City to California to fight in the Mexican War. The men mustered out of the army at war's end, and some of them chose to seek work in California rather than return directly to Salt Lake City. As explained in the introduction, some of these Mormon Battalion veterans were at work digging John Sutter's millrace when James Marshall discovered gold there in January 1848. After trying their hand at gold digging during the spring of 1848, some of the veterans returned to Salt Lake City that summer by blazing what would later be called the Carson Route.

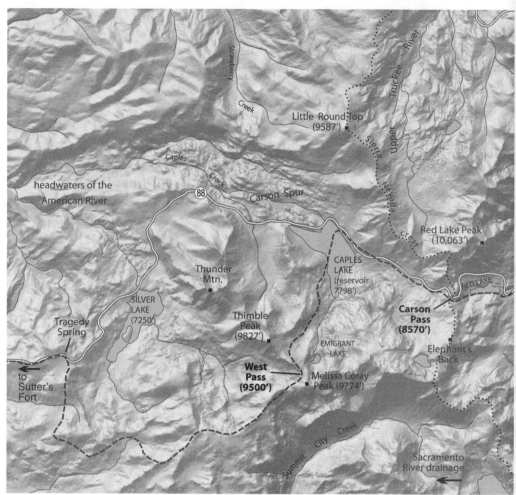

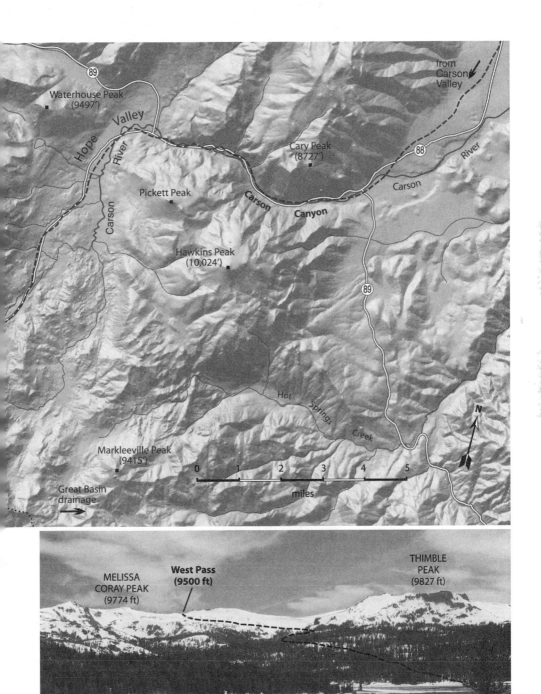

Waterhouse Peak
(9497')

89

from
Carson
Valley

Hope
Valley

Carson
River

Cary Peak
(8727')

88

Carson   River

Pickett Peak

Carson

Canyon

Carson

Hawkins Peak
(10,024')

89

Hot   Springs

Creek

N

Markleeville Peak
(9415')

0   1   2   3   4   5

miles

Great Basin
drainage

MELISSA
CORAY PEAK
(9774 ft)

West Pass
(9500 ft)

THIMBLE
PEAK
(9827 ft)

CAPLES LAKE (7798 ft)

pass. The fiendish scramble was known as the Devil's Ladder—an ascent that Franklin Langworthy deemed "the most dreaded by emigrants of any upon the entire land route to California."

The feat of taking wagons up the Devil's Ladder can only be appreciated by climbing up yourself. The route—thickly studded with pines clinging to cracks in fractured granite—is so steep in places that you can reach out and touch the rock in front of you without bending over. "Had we met such an ascent in the earlier part of the journey," William Kelly wrote, "I fancy we should have pronounced it insurmountable, and turned back in despair." But now, he added, having "overcome so many difficulties, we became inured to hazard and toil, only regarding the greatest obstacles as merely perplexing, but never impossible."

Up they went, using block-and-tackle strapped to trees, multiple teams of oxen, horses, or mules, and all hands pushing and turning the wagon wheels. Going straight up was out of the question. Instead the trail switchbacked up the slope. "The road is crooked, taking numerous short turns around the roots of huge trees, and in some places, is paved over with large roundish rocks," Franklin Langworthy explained. "Up, and over these, the cattle were compelled to climb, sometimes slipping down, and in other instances creeping upward upon their knees." Sometimes gravity won. J. Wesley Jones witnessed a grisly accident in 1851 in which "the Horses [were] unable to hold the wagon at rest, it rolled back with them over a precipice and they fell a mass of fragments at its base."

Gaining the divide at Carson Pass—the split between Pacific-bound and Great Basin–bound streams—you might expect that everyone could have now rolled downhill toward the gold country. But the quirks of glacial erosion wouldn't make it so. A few miles downhill to the west lay a set of impassable crags and cliffs called the Carson Spur.[4] The only way around it was to go up and over yet *another* pass—9,500-foot West Pass (also called Second Summit or Emigrant Pass). After the struggles of Carson Canyon and Carson Pass, the prospect of yet another ascent

---

4. Today's Highway 88, which stays within shouting distance of the emigrant trail all the way from Carson Canyon to Carson Pass, departs from the trail a few miles west of the pass to take a dynamited shortcut along the once-impassable Carson Spur.

13.4 (*previous pages*) Carson Pass area map. The Carson Route across the Sierra Nevada involved three great hurdles: Carson Canyon, Carson Pass, and West Pass. The left photograph looks west from Red Lake up the Devil's Ladder to Carson Pass. The right photograph looks south across Caples Lake to West Pass. The dashed lines show the approximate locations of the trail. Both are June photographs; there would have been less snow when the emigrants came through in August and September.

must have been disheartening in the extreme. But whether they had by now run through their stock of superlatives, or because the climb to West Pass was easier than the ones up Carson Canyon or Carson Pass, the ascent earns comparatively little commentary in emigrant accounts. The trail runs diagonally up yet another cirque wall to a knifelike arête—a narrow, glacier-cut ridge between back-to-back cirques. Ascending, you pass through the tree line—rising like a ragged green wave against the granite slopes—and climb on into a world of glacier-cut rock and perpetual snow.

West Pass is one of the highest emigrant passes over the Sierra Nevada. Indeed, it was the highest place that nearly every emigrant man, woman, or child would stand in their lives. As recently as 12,000 years ago, the crest of the Sierra Nevada lay entombed under groaning rivers of ice. Oozing east and west off the range, the glaciers scooped out countless cirques and carved the range crest into sharp ridges and pointy summits. Small ice streams merged into colossal ice rivers that excavated magnificent valleys—Kings Canyon, Kern Canyon, Yosemite Valley, and others. "Nothing in nature I am sure," James Pritchard wrote at West Pass, "can present Scenery more wild, more rugged more bold, more grand, more romantic, and picturesquely beautiful, than this mountain scenery." James Wilkins agreed. "The scenery is sublime, vastness being the great feature to express in a picture of it. Here on the very summit of the back bone of the American continent, we were favored with a storm of hail, rain, and sleet. The wind blew icy cold." Chill and wind aside, everyone exulted in their accomplishment. "The Summit is crossed!" Niles Searls cheered on October 1, 1849. "We are in California! Far away in the haze the dim outlines of the Sacramento Valley are discernable! We are on the down road now, and our famished animals may pull us through."

THE EMIGRANTS' troubles weren't over once they had crested the Sierra Nevada passes. No matter what trail they were on, they still faced 50-plus miles of hard hauling down the mountain's western slope. "We had expected an easier road down the mountains after crossing the main ridge, but were disappointed," Margaret Frink grumbled. For the most part, the trails down the western slope run along the tops of ridges between glacier-cut and river-cut valleys. The ridgetops are the remains of a broad riverine plain that existed before the Sierra block began to rise and tilt west (fig. 12.8). Their smooth tops thus form natural avenues for east-west travel—far better than the rough, eroded valleys in between. (Many of today's east-west highways follow the ridgetops.) Where ridgetop travel was not an option, the emigrants had to drop down into the valleys, sometimes down slopes so steep that they lowered their wagons on ropes.

One way or another, the work was done and within a few days they rolled—tired, eager, and happy—into the land of gold. Now, after 2,000 miles of hard roads, they were emigrants no more. They would be miners now—or merchants or hoteliers or mechanics or (for those who opted for land over gold) farmers or ranchers. "We here saw for the first time the process of gold working a going on," Joseph Hackney wrote excitedly as he came to his first diggings on September 11, 1849. "We all here got out our pans and went at it and washed out a half-dollar's worth in no time. The first money I ever made out of the land."

JOSEPH HACKNEY'S first taste of gold made it seem easy. But for most who chased the yellow metal, California would yield only poverty and disappointment. Alonzo Delano,[5] who traveled far and wide through the Sierra Nevada during the gold rush, offered this assessment:

> Wherever we turned, we met with disappointed and disheartened men, and the trails and mountains were alive with those whose hopes had been blasted, whose fortunes had been wrecked, and who now, with empty pockets and weary limbs, were searching for new diggings, or for employment—hoping to get enough to live on, if nothing more. Some succeeded, but hundreds, after months and years of toil, still found themselves pining for their homes, in misery and want, and with a dimmed eye and broken hopes.

In the 10 years that followed the rush of '49, miners sifted, scraped, and blasted 28 million ounces of gold out of the Sierra Nevada foothills. It was by far the greatest concentrated gold strike in human history up to that point, worth $594 million at the time (more than $10 billion today). But when you consider that more than 100,000 miners were in on the competition during those 10 years, that works out to less than $600 per man per year. In the sky-high prices of gold rush California, simply feeding yourself could cost that much.[6] Moreover, the biggest strikes were made by a comparative few. Many more wandered from one barren digging to another, sometimes for years, growing ever more impoverished and disillusioned. "Even in the most auriferous sections," Alonzo Delano explained, "there is only a comparatively small portion which pays the

5. We last heard from Delano along the Humboldt River. He turned off onto the Applegate-Lassen Trail at Lassen Meadows, opting for the northern route around the Sierra Nevada rather than the more direct Truckee or Carson routes.

6. William Swain thought he was doing well if his cost of board dropped to $1.50 per day (Holliday 1981, 359). During shortages, food prices skyrocketed. As one miner near Mariposa explained, "The price of provisions had become so high that our paltry earnings were not nearly sufficient to pay for the food we required to keep us alive."

laborer abundantly; and while now and then one miner may make a good strike, by far the greater number will make scarcely day wages." Seeing the occasional big strike only compounded the anguish. "Oh Caroline," one tormented miner wrote to his wife, "I can't bear the idea of going home with so little money as I have now got when there is a fortune so near." "Say to all my friends: stay at home. Tell my enemies to come [to California]," wrote another embittered soul. "I really hope that no one will be deterred on coming," another ruefully proclaimed. "The more fools the better, the fewer to laugh when we get back home."

∗

Thank—or blame, if you prefer—geology for the gold rush. We began our journey with gold's arrival on planet Earth.[7] We'll end it by considering how such a sizable fraction of the Earth's gold ended up in California. The answer—it turns out—arises from a three-way geologic convergence of ancient rivers from the east, ancient seabed from the west, and granite magmas from below.

A billion atoms selected at random from the Earth's crust will yield, on average, about five atoms of gold. To appreciate what five parts per billion means, imagine a football field covered with a billion pieces of green paper, five of which are $100 dollar bills. Easy money? Think again. Assuming it takes one second to inspect each piece of paper, you could be on that football field for up to *31 years* before you made $500. The point is, gold mining could not exist if geologic processes didn't somehow gather up that five parts per billion of gold and collect it together into super-abundant concentrations. In California that happened in two main ways—by the erosion of gold out of bedrock and by the formation of gold-bearing quartz veins.

Gold is soft and easily eroded, but it resists chemical attack. It is also heavy—nearly twice as dense as lead. These properties—soft, stable, and heavy—account for placer gold—gold eroded from bedrock sources and concentrated in streambeds. (Placer—pronounced like "passer," not "pacer"—is a Spanish nautical term for sandbank.) Eroded out of bedrock veins and washed downhill, heavy gold particles collect on the bottoms of stream channels or in the eddy zones downstream of boulders or ledges.

Streambed placer mining during the California gold rush crested and crashed with giddying speed. The peak year was 1852, with nearly

---

7. As explained in the introduction, the Earth (according to current theory) is the coalesced remains of stardust—the elemental residue of ancient exploded stars. That residue includes some 100 known chemical elements, including gold.

13.5 Nuggets of placer gold on display behind bulletproof glass at a casino in Carson City, Nevada. The container is about two inches tall.

4 million ounces mined. As the easy placers were mined out, production declined rapidly (fig. 13.6). By the Civil War, California was producing less than 1 million ounces of gold per year, and production since then has rarely risen above that mark, despite vast improvements in mining technology.

Even before the streambed placers dried up, miners were looking up-hill for the sources of all that streambed gold. They soon realized that there were two: fossil rivers—ancient, uplifted streambeds now eroding and shedding gold into living streams—and gold-quartz veins—a web-work of opaque whitish seams, riddled with gold, that shoot through the bedrock of the western Sierra Nevada like cracks in shattered glass.

Along Interstate 80 near Gold Run, 3,200 feet above sea level in the western Sierra Nevada, the highway sweeps past a stunning road-cut, nearly 100 feet high and more than a quarter-mile long, stained rust-red and orange. Closer inspection shows that the outcrop is packed with mil-lions of baseball-sized cobbles that fill ancient river channels—gigantic ones, as if several Susquehannas or Colorados had been stacked on top of one another there in the mountains. To see these rivers in their heyday, spin the clock back 50 million years. You now hang suspended some 3,000 feet above the ground. There is no Sierra Nevada range under your feet—it doesn't yet exist. Instead, lush, forested lowlands sprawl below you, bisected by huge rivers flowing west. Far to the east, in Nevada,

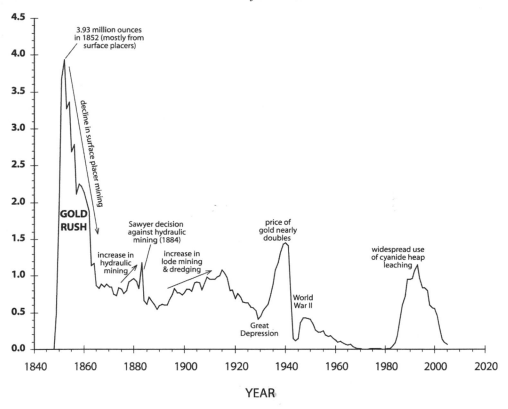

13.6 The history of gold production in California from 1848 to the present. The great peak centered on 1852 was almost entirely from surface placers. Subsequent production variously involved hydraulic mining of the auriferous gravels (until 1884), lode mining, and large-scale dredging of riverbeds from barges. The most recent surge in production, beginning in the 1980s, involved not new gold discoveries but a new technology— cyanide heap leaching—that allowed gold to be profitably extracted from previously unprofitable sources.

lie snowcapped mountains—highlands left over from the Sevier Orogeny—from which the rivers run. There the rivers tumble through gorges riddled with gold. Their churning waters are collecting Nevada gold and shipping it west.[8]

Fast-forward now to 5 million years ago. The great block of crust that will become the Sierra Nevada range begins to lever upward and tilt west. The rising mountain jacks the gold-bearing fossil riverbeds into the sky.

8. Nevada has always outranked California as a golden state. The emigrants unknowingly passed by far more gold as they hurried across Nevada than would ever be mined from California. Put California's total gold production in a pile, and it would be outweighed by Nevada's total production from just the last four decades. Most of it comes from mines along the Carlin Trend in north-central Nevada.

New streams establish themselves on the mountain, slicing down as the mountain goes up. The streams cut down through the fossil riverbeds as they carve out their valleys, capturing much of the gold in the process. Fossil riverbeds that escape erosion are left high and dry on valley walls and intervening ridges. Fast-forward again, to the gold rush years. Miners look up from their dwindling streambed claims. They realize that vast, untapped riverbeds lie overhead—dozens of them, fossilized and silent, winding through 8,000 square miles of the western Sierra Nevada. They call them the auriferous gravels—*auri* for gold and *ferous* for their signature iron-oxide blush. Their discovery will kick off a new era in gold rush history—the era of hydraulic mining.

The loose gravel of a babbling Sierra Nevada brook can be washed well enough by funneling some of the stream through a sluice. But you can't wash a fossil Colorado River that way—especially one stranded on a sun-blasted ridge miles from the nearest living stream. To attack the auriferous gravels, gold rush miners had to come up with new weaponry. The solution (pioneered in 1853 by Connecticut miner Edward Matteson) was to impound streams high up in the mountains and funnel the water through miles of ditches and flumes to points *above* the fossil riverbeds. Once the water arrived on site, miners sent it steeply downhill through riveted steel pipes and canvas hoses. At the ends of the hoses, they installed swivel-mounted water cannons, called monitors, with mouths the size of a man's head or larger. The pressurized water blasted out at more than 100 miles an hour—powerful enough to take off an arm. In scenes reminiscent of modern high-rise firefighting, the miners directed these titanic sprays against the auriferous valley walls—and the ancient riverbeds crumbled before the onslaught.

It was a brute-force approach, and probably quite inefficient in terms of the fraction of gold recovered. But hydraulic miners could wash so much riverbed gravel so quickly that the method still netted huge sums of gold. It also spawned ecological disaster. As hydraulic mining exploded throughout the western Sierra Nevada from the late 1850s through the early 1880s, millions of gallons of muddy slurry gushed into the valleys of the Feather, Yuba, Bear, and American rivers. The clogged rivers rose in debris-laden floods to drown farms and ranches across the Central Valley. Along the Yuba River alone, 18,000 acres of farmland were smothered in mining effluvia. The debris moved on to the Sacramento River, where it gathered into giant sandbars that threatened steamboat navigation. Mud and silt laid waste to the deltas at the head of San Francisco Bay and stained the water brown all the way to the Golden Gate. The roar of the monitors drowned out the voices of protest for decades. But in 1884 a citizen's group called the Anti-Debris Association finally won

13.7 Hydraulic mining in 1869 at the Malakoff pit near North Bloomfield. The scene shows a small portion of a gaping, man-made valley that eventually grew to be 7,000 feet long, 3,000 feet wide, and 600 feet deep, all blasted out by high-pressure water. By the time mining ended here in 1884, miners had removed an estimated 4.2 billion cubic feet of gravel (enough to fill about 80 football stadiums to the brim) and recovered $3.5 million in gold. The insert photograph shows a monitor—the water cannons used to tear down the gravel beds. Carleton E. Watkins, *Malakoff Diggins, North Bloomfield, Nevada County, Cal.*, ca. 1869, mammoth plate albumen print, 15¾ × 20⅝ in., San Francisco Museum of Modern Art. Purchased through a gift of the Judy Kay Memorial Fund.

a legal fight against the hydraulic miners. The court decision banned the flushing of hydraulic mining debris into the rivers of the western Sierra Nevada. While the decision did not outlaw hydraulicking per se (miners could still hose down mountainsides, as long as they caught the runoff with dams and debris basins), it effectively ended the practice by making it unprofitable.

ULTIMATELY, ALL placer gold—whether from living rivers or fossil ones—requires a bedrock source. In California, quartz veins are that source. "Miners are in the habit of remarking that, where there are good quartz veins there are almost invariably good placer diggings," Alonzo Delano observed. You can find quartz veins in most road-cuts and outcrops in the western Sierra Nevada foothills. They appear as crystalline white seams, a fraction of an inch to several inches thick (although

13.8 Quartz veins cutting through the Calaveras Complex along the American River. Note the quarter coin in the foreground for scale.

large ones may be more than 100 feet thick). Theory holds that quartz veins form by hydrothermal precipitation—the crystallization of dissolved quartz and precious metals out of hot underground waters. Where pluton-forming magmas rise to within a few miles of the Earth's surface, they meet groundwater. The water, superheated by passing through or near the magma, dissolves just about anything it finds. It picks up trace metals from the magma and the surrounding rock—iron, sulfur, lead, magnesium, copper, and gold. The water also dissolves huge quantities of silica. (Silica [SiO_2] is the most abundant chemical compound in the rock of the Earth's crust.) The hot, silica-rich water squirts upward along faults and fractures. As it cools, the silica crystallizes out as milky-white quartz, harder than steel. Metals like gold—if present—precipitate as well, entombed in quartz.

The gold content of quartz veins is maddeningly unpredictable. Pure quartz veins are typically white. But gold-bearing quartz veins often contain iron sulfides and other impurities that stain them red, brown, or yellow with promise—or false hope, because color doesn't guarantee the presence of gold.[9] There is no way to tell just by looking at a quartz vein whether it will make you a prince or a pauper. You must dig or blast the quartz out of the bedrock, crush it to release the microscopic flecks of gold, and then either bind up the gold with mercury or dissolve it with cyanide solution. You hope for riches; you pray to meet expenses.

Quartz veins were called lodes by gold rush miners, and it wasn't long after the rush of '49 that miners discovered an astonishingly rich belt of lodes, scarcely two miles wide but 150 miles long, running north-south through the western Sierra Nevada foothills. Mining camp after mining camp sprang up along this lode belt as miners gophered their way into the quartz-rich rock. It was truly the mother of all lodes, and they called it the Mother Lode.

You can trace the Mother Lode belt on any California map. Just find Highway 49. This quiet, twisting foothill road follows the Mother Lode closely, and the towns and historic markers along the way read like the pages of gold rush history: Chinese Camp, Columbia, Carson Hill (where in 1854 miners dug up the largest nugget ever found in North America: 195 pounds), Angels Camp, Mokelumne Hill, Jackson, El Dorado, Hangtown (now Placerville), and Gold Hill.

For me, Highway 49—the Mother Lode highway—is much more than a gold rush tourist road. It is a geo-historical lens. Highway 49 focuses two parallel stories—one of a continent, the other of a country—down onto a single crystalline point.

To see what I mean, we'll hit the highway they call Highway 49 in Late Jurassic time, 150 million years ago. We're not hearing the wind in the foothill oaks now, and we're not smelling the equine funk of nearby ranches either. Instead, we're hearing ocean waves and smelling salt

---

9. According to Alonzo Delano, "The best veins abound in pyrites, or sulphuret of iron, and the best quality of ore is of a dull brownish red, which varies in its shade, being colored with oxide of iron."

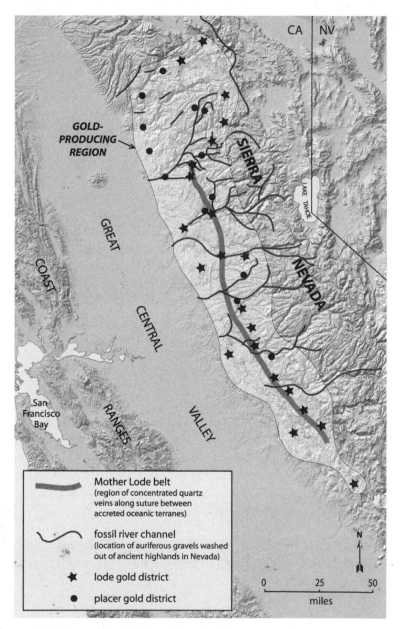

13.9 California owes much of its gold production to two main sources: (1) the gold-bearing auriferous gravels—fossil riverbeds uplifted and eroded as the Sierra Nevada rose; and (2) the Mother Lode belt—a band of quartz veins formed along an ancient collision zone between two oceanic terranes. Underground hydrothermal fluids took advantage of the fractured rock of the collision zone to percolate upward and deposit gold-quartz veins (lodes). Lode districts that are not on the Mother Lode belt occur along other fault zones that likewise became sites of hydrothermal precipitation.

spray. West of the highway, a dinosaur-footprinted beach catches Pacific surf, and ichthyosaurs leap beyond the breakers.

California, remember, is mostly an assemblage of oceanic terranes—slabs of ancient seabed scraped off the Farallon Plate (chapter 7). The Mother Lode belt and Highway 49 follow the collision zone between two of these terranes—the glue-line where one terrane crunched into another one in the process of assembling California. The rock east of the highway generally belongs to the older terrane—the earlier one to land on California's then-western edge—while the rock west of the highway belongs to one of several younger terranes.[10] As yet more terranes stacked on farther west, California grew outward in successive increments. But 150 million years ago, California ended at Highway 49. There was no land to the west—no Sierra Nevada foothills, no Central Valley, no Coast Ranges. They hadn't arrived yet.

Back to Highway 49, Jurassic version. We stand on the edge of the continent, looking out across the ancient Pacific. Deep below the rolling swells, the abyssal seabed of the Farallon Plate is sliding east and bending down into the Farallon Trench. It looks like subduction-as-usual—an oceanic plate sinking to its fate, on its way to feed magma to volcanoes in Nevada. But no. A Taiwan-sized slab of seabed is putting up a fight. It doesn't want to become Nevada volcanoes. Instead of sinking into the mantle, it detaches from the Farallon Plate and scrapes off—with millennial slowness—against the edge of the continent. Along the collision zone, the rock shears into a cacophony of faults and fractures that practically invites invasion by quartz veins. The collision zone[11] will become the Mother Lode—for quartz veins follow fractures, and gold follows quartz.

Now we have to wait a bit. To get gold-bearing quartz veins to invade the shattered rock of the collision zone, we need heat—and lots of it—to cook up underground hydrothermal fluids. By mid-Cretaceous time, 120 million years ago, we get the heat. Blobs of magma rise from the mantle above the still-subducting Farallon Plate. The molten masses invade the oceanic terranes, melting their way through untold cubic miles of accreted seabed. Underground waters, heated to temperatures

10. Calaveras Complex is the name of the older terrane east of the highway. Pieces of the younger terrane west of the highway are called variously the Foothills Terrane, Western Jurassic Terrane, or Smartville Complex, depending where you are.

11. California geologic maps identify this ancient terrane collision zone as the Melones Fault and mark it as a black line running close to Highway 49. Given California's shaky reputation, one might reasonably ask whether the Melones Fault is an active earthquake maker. It seems not. The fault appears to have shaken out its last earthquake about 145 million years ago, as the docking terrane to the west crunched to a halt.

far higher than their surface boiling point, circulate through the magma and the terrane rocks. They gather up widely disseminated gold and vast amounts of silica. The hot waters rise buoyantly along the path of least resistance—the fractured terrane collision zone. There they cool, precipitating their cargo of quartz and gold to make the Mother Lode. The magmas cool too, eventually crystallizing into granite plutons. We've seen these plutons before. Together, they constitute the greatest mass of granite in North America—the Sierra Nevada Batholith. Gold rush miners knew that where you find granite, you will find quartz veins. "Quartz and granite appear to hold companionship," Alonzo Delano observed in 1849, although he didn't know why. Hydrothermal waters, we now understand, form the link between granite, quartz, and gold.

Thus, the same forces that assembled our continent, piece by piece, and built the foundation for one its mightiest mountain ranges, also planted the riches that launched thousands onto the hard road west.

* 14 *

CONTINGENT HISTORY

For most California-bound emigrants, the destination was the point. Two thousand miles of hard road was the price to pay for gold, land, and a new life. For those of us who follow them—from our armchairs or along backcountry roads—the journey is the thing.

We look back on that journey from a world transformed. Roads and rails now crisscross landscapes that our gold rush era forebears saw as howling wilderness. The golden arches of McDonald's today rise within shouting distance of the Humboldt River. You can order a piece of cow there from a kid whose ancestors shot arrows into emigrant cows not far away. Skiers at Donner Pass slalom down old wagon trails. Tourists at Donner Lake munch sandwiches where members of the Donner Party spent the winter not munching sandwiches. Yesterday's hardscrabble gold rush towns are today's tourist traps, crowded with T-shirt shops and vacation families.

Yet there are vast stretches along the overland trails where modernity, thus far, has touched lightly. Wyoming has most of them: the Sweetwater Valley, the South Pass area, and much of the Green River Basin. The largest empty sector lies between Idaho's City of Rocks and the head of the Humboldt River—100 miles of dust and sagebrush where asphalt remains a stranger. In these quiet places, you can still can sidle up close to America's past—as close as you're likely to get.

In truth, we can never know the West that the emigrants knew. We know a different West today—several, in fact. There's the casinoed, bill-boarded, barbwired West—the one we could do without. But there's also the West of yawning calderas and boiling springs; of young mountains born of old seabeds; of a continent stitched together yet tearing apart; and of rivers that bury mountains and then dig them back out again. This West has more to teach us—more than I've laid out between these pages. Rock hammers flashing from distant ridges are already uncovering new stories, and campfire geologists, flapping like one-winged chickens (the other hand has to hold the beer), are fanning the flames of new ideas on a western frontier that will never close.

THE HARDSHIPS and triumphs of the emigrant journey unfolded as they did because of contingent events buried in the depths of time. Historians like to talk about contingency—the notion that key events in the past (turning points, if you like) determine the course of subsequent history. "Contingency is central to any historical process," the historian David Hackett Fischer tells us. Consider the ill-fated charge of a Civil War general named Pickett. Viewed in isolation, Robert E. Lee's decision to order the charge at Gettysburg seems nuts. Send men across a mile-wide open field straight at the muzzles of ready Union guns? But consider contingency. Just two months earlier, at Chancellorsville, Lee had audaciously attacked an immensely superior Union force and won. Lee's success at Chancellorsville was arguably the contingent event that made him overreach at Gettysburg. The eventual collapse of the Confederacy can itself be traced back through myriad contingencies. One of these, as I'll argue shortly, was the California gold rush.

Historical contingencies don't always fit into the scope of human time. This book's central argument is that North America's geologic history—the multimillion-year history of a continent heading west—guided the course of America's own history of westward migration. Several major themes of geo-historical contingency emerge from this journey.

First—and of singular importance—is North America's long history of east-west plate convergence. The North American Plate headed west while the Farallon Plate plunged under it to the east. The result is a landscape first squeezed and later stretched east-west, so that the mountain ranges and valleys tend to run north-south. Rotate the mountains of the West in your mind so that they line up mostly east-west, and think of how much easier it would have been to travel overland, following the trend of ridges and valleys instead of slashing across them or finding routes around.

The Rocky Mountain Foreland Ranges are one result of this long history of plate convergence. They reared up astonishingly far inland of North America's western edge—a consequence, we suspect, of the Farallon Plate subducting flat beneath the continent for a time. The upshot of this contingency was that westbound emigrants ran into mountains while still 1,300 trail miles from California or Oregon. More recently, the burial and subsequent exhumation of the Foreland Ranges allowed rivers to cut down through once-buried ridges via superposition. For the emigrants, this created both physical barriers (impassable canyons) and puzzles (Devils Gate on the Sweetwater River). Finally, the collapse of one of these ranges (the Sweetwater Range) opened the Sweetwater Valley route to South Pass—the great funnel through which the westward migration poured.

Farther west, the emigrants entered the Basin and Range and the Great Basin. When North America's westward migration finally snuffed out the Farallon Plate, compressive mountain building came to an end, and hot mantle rock welled up below much of the Southwest. In response, the crust stretched some 250 miles east-west to make the north-south mountains of the Basin and Range, while sagging in the center to form the Great Basin. This contingency condemned the emigrants to the route of the Humboldt River—the only stream that flows west for any appreciable distance through the north-south grain of the Basin and Range landscape. It also meant that they faced a river-less leap at the end, because all Great Basin rivers die in the Great Basin. That leap came at either the Forty-Mile Desert or (for those who opted for the Applegate-Lassen Trail) the Black Rock Desert.

Closing in on California, we encounter the most significant geo-historical contingency of the journey—the uplift of the Sierra Nevada. The rising range brought aridity down on the Great Basin. That aridity, together with summer's heat, served up the most brutal experiences of the journey as the emigrants rolled west along the shrinking, salinized Humboldt, and then stepped off into the river-less desert beyond. Then, pulling into the sunset shadows of the Sierra Nevada itself, they confronted the range's eastern escarpment—their most daunting barrier.

Finally, we have the gold itself. A geologic trinity brought gold to California: seabed terranes collided from the west; gold-bearing rivers washed down from the east; and granite-forming magmas welled up from below. The results were streambed placers, auriferous gravels, and gold-quartz veins—the bait of the gold rush. Arguably, no geo-historical contingency has reached further through American history than the formation of California's gold. It did more than trigger a great migration. It propelled a backwater territory to statehood practically overnight and may even have saved the Union. Shipped east in the years up to and during the Civil War, the yellow metal poured mostly into the coffers of northern banks and the federal treasury. (The northern states had the big financial institutions and a manufacturing industry that rewarded capital investment. The southern states—agrarian, wedded to slave labor, with little manufacturing infrastructure—got little of the gold.) California gold became Union bullets, boots, rifles, cannons, uniforms, locomotives, and railroads. Robert E. Lee acknowledged this, indirectly, in April 1865 when he told his weeping soldiers at Appomattox that the Confederate army had "been compelled to yield to overwhelming numbers and resources." California emigrant John Bidwell, reflecting back on the war years, summed up the significance of California's gold this way:

> It is a question whether the United States could have stood the shock of the great rebellion of 1861 had the California gold discovery not been made. Bankers and business men of New York in 1864 did not hesitate to admit that but for the gold of California, which monthly poured its five or six millions into that financial center, the bottom would have dropped out of everything. These timely arrivals so strengthened the nerves of trade and stimulated business as to enable the Government to sell its bonds at a time when its credit was its life-blood and the main reliance by which to feed, clothe and maintain its armies. . . . The hand of Providence so plainly seen in the discovery of gold is no less manifest in the time chosen for its accomplishment.

EPILOGUE

Alonzo Delano has little luck with gold. He hangs up his pick and shovel and becomes a merchant and banker. By 1851 he has done well enough to bring his family to California. They settle in Grass Valley, on the Mother Lode belt. Modest fame comes with a series of humorous newspaper articles about miners and mining, and with the publication of his 1854 book chronicling his overland crossing and adventures in the gold country. He always writes affectionately of his fellow miners—those hardworking men "delving among the mountains, hoping to acquire a competence to smooth the down-hill of life, and render old age comfortable."

Margaret Frink and her husband, Ledyard, try a bit of gold panning, but they suspect that surer money lies in the growing bustle of Sacramento. They survive a deadly cholera outbreak that wracks the city in October 1850 and prosper thereafter as hoteliers. With natural business acumen, they figure out what people want but can't easily get—for instance, offering free fresh milk to all guests, "a great attraction to men, many of whom had not tasted milk for one or two years." They prosper and grow wealthy. "The progress of time only confirmed to us more strongly our choice of a home," Margaret concludes in her journal, "and we never had occasion to regret the prolonged hardships of the toilsome journey that had its happy ending for us in this fair land of California."

While thousands of gold rush emigrants—like Alonzo Delano and Margaret Frink—put down permanent roots in California, thousands of others yearn only to return home. William Swain is one. "I am coming back again with a pocketful of rocks!" he vows to his wife, Sabrina, upon

This 1869 photograph dramatically symbolizes a tipping point in American history. Westbound wagons pass by the famous Jupiter train heading east, carrying Leland Stanford to the Golden Spike ceremony in Utah that will unite the continent by rail—and end the era of wagon train emigration. Used by permission, Utah State Historical Society, all rights reserved.

setting out west in 1849. After one backbreaking year at the mines, his net take is about $500. Homesick, and despairing of his failure to amass a fortune, he returns in 1851 (by Panama steamer) to his wife and baby daughter in western New York State. He farms there for the rest of his days.

Like William Swain, Byron McKinstry wants only to get as much gold as he can out of California and go home. He makes the overland crossing in 1850 and spends 21 months diligently washing gravel from gulches near Mokelumne Hill. He does better than many, returning home to Illinois in 1852 with $1,315 in gold and cash, which he immediately invests in some rich farmland south of the young city of Chicago. There he prospers raising fruit trees and cattle. In 1883 he suddenly announces to his wife that he is leaving for the Dakota Territory to try homesteading. He fails miserably in that endeavor and returns home ten years later a broken and lonely man.

A. J. McCall sets out across the plains in 1849 with "no faith in the marvelous tales of golden nuggets shimmering in the mountain streams of the Pacific slope." Perhaps because of low expectations, he has a rip-roaring good time hunting for gold. "The last six weeks have been the happiest of my life," he declares of his time picking and panning with friends along the Yuba River. The rough, independent life of a miner seems to have agreed with McCall—as it probably did with many men. Despair and disappointment abounded in the California mines—no question—but there was also fun and wild freedom. As Mark Twain wrote in 1871, the California miners "fairly reveled in gold, whiskey, fights and fandangos. . . . It was a wild, free, disorderly, grotesque society!" After scraping together about $300 worth of Yuba River gold, McCall leaves the mines in late 1849 and moves to Sacramento, where he lives for about one year before returning home permanently to New York State.

J. Goldsborough Bruff leads a train of forty-niners to California by the Lassen Route, only to be abandoned by them in the northern Sierra Nevada when he becomes too sick to travel. Alone with his dog over the winter and too weak to move, he nearly starves. He eats dead birds, boils deer bones, and dreams of shooting an Indian for food. "My mouth fairly watered, for a piece of an Indian to broil!" Rescued in the spring of 1850, he finally reaches the gold country and spends two years prospecting—all for naught. Eventually he returns to Washington, D.C. (via Panama), where he resumes his pre-gold rush position as a draftsman and architect for the U.S. Treasury Department. He doesn't consider his California interlude a complete failure, though, for as he explains, "I had 'seen the elephant,' and emphatically realized the meaning of the ancient myth—traveling in search of THE GOLDEN FLEECE!"

Edwin Bryant makes his first crossing to California in 1846 and—with impeccable timing—publishes *What I Saw in California* in 1848, shortly after the news of the gold discovery at Sutter's Mill bursts upon the world. The book becomes an instant best seller among gold rush emigrants and armchair adventurers alike. Bryant returns overland to Kentucky in 1847, and then crosses again to California in 1849 at the head of a party of gold-seekers. He is by now a folk hero, and several forty-niners note seeing the famous pioneer on the trail. Arriving in San Francisco for a second time, he sells for $100,000 some waterfront property he had bought for $4,000 on his first visit two years before. Wealthy now, he returns home to Kentucky (via Panama), having reaped a fortune from California without ever having swung a pick or swirled a pan. In 1869—aging and frail—he boards the new transcontinental railroad for his final overland crossing. From the train, he watches the old emigrant trail clip by at an astonishing 20 miles per hour and ponders the closing of a chapter of history that he helped to write.

ACKNOWLEDGMENTS

Researching and writing *Hard Road West* was more work and more fun than I ever thought possible. Converting a 2,000-mile-long camping trip into something worth putting between book covers took time, and not just from me. The process went well because of the people who helped — colleagues, friends, and family members who tried, unsuccessfully, to hide when they saw me coming with pages flapping. My early stabs at a book proposal eventually became something contract-worthy thanks to critical input from Susan Brown and Jim Malusa, and to my agent Susan Rabiner, whose book *Thinking Like Your Editor* should be required reading for every aspiring author. Christie Henry, my editor at Chicago, provided skillful guidance and useful feedback from the moment my proposal crossed her desk. Erin DeWitt provided masterful copyediting and fixed numerous errors and omissions. My parents (to whom I dedicate the book), along with Betsy Wieseman and Allen Wright, helped clean up much of my writing. Malcolm Meldahl and Jim Walsh read every chapter with ruthless editorial eyes, gunning down clichés and kicking me back on track with helpful comments like "ugh," "no!" and "get rid of this." There aren't enough good cigars on the planet for these two fine men. Richard Rieck, bringing expertise in both geology and trail history, read the manuscript and wrote a sizable one of his own in response, fixing many errors of fact and usage. Three anonymous reviewers — a professional geologist, a geographer, and a historian — improved the manuscript with many insightful comments. Alan Cutler shared his expertise on the history of geologic thought. Charles Martin Jr. shared his insights into the geologic perspectives of the emigrants. Will Bagley supplied several key emigrant quotations. Tom Wilcockson of Mapcraft, Inc., produced the two introductory trail maps.

I completed the bulk of my travel and research while on sabbatical from Mira Costa College. I thank my friends and colleagues on campus for their enthusiastic support of this work. During my sabbatical, the Department of Earth Sciences at the University of California at Santa Cruz kindly granted me a visiting faculty position, thus opening for me the prodigious resources of the University of California library.

I thank the many geologists whose published works have informed the science presented here. This project took me across the heart of the North American Cordillera—home to some of the most complex and fascinating geology on Earth. Many of the rocks, rivers, and mountains were old acquaintances from my rambles across the West. But more were new. The work steered me into topics and regions far from my experience, and I was humbled—and thrilled—to gaze upon outcrops that confused me in ways I haven't felt since my undergraduate days. I've tried to get everything right; I hope I've mostly succeeded.

I thank the Oregon-California Trails Association and Trails West, Inc., for their efforts at preserving and protecting America's historic overland trails. Because of their efforts, you and I can still stand in the traces of the westward wagon ruts and touch history. And I thank the emigrants of the gold rush era for leaving us their accounts of the overland crossing—a journey that was, I believe, America's greatest adventure.

I thank my wife, Susan, for love and companionship; my friend Dan for delightful desert debauchery; and the rest of my family—my father, Edward, along with Joyce, Virginia, Malcolm, Joseph, Geoffrey, and Ethan—for their unflagging support. Most of all, I thank my mother, Eleanor, who raised me with core values of hard work, economy, curiosity, and love for ideas. She, more than anyone, has made me who I am.

NOTES

To introduce myself to the geology along the trails, I used the state geologic maps of Burchett 1986 (Nebraska), Love and Christiansen 1985 (Wyoming), Bond and Wood 1978 (Idaho), and Stewart and Carlson 1978 (Nevada), as well as the *Roadside Geology* guidebooks of Maher, Englemann, and Shuster 2003 (Nebraska), Lageson and Spearing 1991 (Wyoming), Alt and Hyndman 1989 (Idaho), and Alt and Hyndman 2000 (California).

My main sources of geologic information were professional papers published in peer-reviewed journals and edited volumes. These sources are listed in the chapter-by-chapter notes below. Most of this material is technical and not particularly user-friendly for the nonspecialist. The bibliography does, however, include some sources aimed at nonspecialists. I recommend the *Roadside Geology* guidebooks for a good introduction to the geology along the trails, and McPhee 1998 for wonderful writing about the geology the American West.

For the history of westward migration before and during the California gold rush, my main sources were Brands 2002, Holliday 1999, Holliday 1981, McLynn 2002, Stewart 1962, and Unruh 1979. I also drew upon interpretive materials provided at historic sites along the Oregon-California Trail by various state agencies, the National Park Service, and the Bureau of Land Management. These agencies also maintain useful trail history Internet sites, as does the Oregon-California Trails Association.

For firsthand information about the overland journey, I relied mostly on the emigrants' own accounts. Particularly helpful were the journals of Edwin Bryant (Bryant 1848), J. Goldsborough Bruff (Read and Gaines 1949), Alonzo Delano (Delano 1854), Margaret Frink (Holmes 1983, vol. 2), A. J. McCall (McCall 1882), Byron McKinstry (McKinstry 1975), and William Swain (Holliday 1981). Several secondary sources also gave me troves of emigrant quotations and historical information, particularly Holliday 1981, Willoughby 2003, Brock 2000 (for the Humboldt River segment), Curran 1982 (for the Humboldt River and the Forty-Mile Desert), and Fey, King, and Lepisto 2002 (for the Forty-Mile Desert and Sierra Nevada).

To locate historical markers and trail segments along the Oregon-California Trail, I used the guidebooks of Brock 2000, Fanselow 2001, Fey, King, and Lepisto 2002, Franzwa 1988, Franzwa 1999, and Tortorich 2002.

The quotation sources below list the author's last name (or abbreviation) and page number, with the full citation then given in the bibliography. For any source not listed in the bibliography, I provide a full citation along with the quote.

Abbreviations used:

B:	Brock 2000
Br:	Bryant 1848
C:	Curran 1982
D:	Delano 1854
FKL:	Fey, King, and Lepisto 2002
H:	Holliday 1981
Ho1:	Holmes 1983, vol. 1
Ho2:	Holmes 1983, vol. 2
Ho3:	Holmes 1983, vol. 3
Ho4:	Holmes 1983, vol. 4
M:	Morgan 1959
Mc:	McCall 1882
McK:	McKinstry 1975
NF:	National Frontier Trails Museum 2004
R:	Royce 1932
RG:	Read and Gaines 1949
S:	Scamehorn 1965
W:	Willoughby 2003

EPIGRAPH

vi   *"History is all explained by geography."* Robert Penn Warren, in *Talking with Robert Penn Warren*, edited by F. C. Watkins, J. T. Hiers, and M. L. Weaks (Athens: University of Georgia Press, 1990), 26.

vi   *"Any man who makes . . . "* Alonzo Delano, October 12, 1849: H, 175.

vi   *"One only hope sustains . . . "* Margaret Frink, August 20, 1850: Ho2, 143–44.

PREFACE

xiv   *"[I have] undergone more hardship . . . "* William Wells, 1849: Unruh 1979, 414.

xv   *"My God, McKinstry, why . . . "* Byron McKinstry, 1850: McK, 17.

xvi   *"Eastward I go only by . . . "* and *"I must walk toward Oregon . . . "* Henry David Thoreau, "Walking," *Atlantic Monthly*, June 1862.

INTRODUCTION: STARDUST

Information on stellar synthesis of the chemical elements is from Tyson, Liu, and Irion 2000, 61–103.

Kirkemo, Newman, and Ashley 2006 provide information on the properties of gold.

The account of James Marshall's gold discovery is from Marshall's 1857 letter, and from Brands 2002, 1-19.

xix   *"Having some general knowledge . . . "* James Marshall, 1857, 200.

xx   *"If it is gold, it will . . . "* and *"there was my gold . . . "* Jennie Wimmer, 1874: Levy 1990, xx-xxi.

xx   *"it triggered the most . . . "* H. W. Brands 2002, 24.

CHAPTER 1: AN AMERICAN JOURNEY

The story of Sarah Royce's crossing of the Forty-Mile Desert and Sierra Nevada is from her memoir: Royce 1932, 33-75. All of the geology introduced in this chapter is covered in more detail in later chapters; see those chapter notes for specific sources.

1   *"As when some carcass . . . "* Hubert Howe Bancroft, 1884: Brands 2002, 23.

1   *"Turn back! What a chill . . . "* R, 45.

2   *"So you've given out . . . "* R, 52.

2   *"scenes of ruin . . . "* R, 53.

2   *"So faithful had they been . . . "* R, 56.

2   *"Was it a cloud? . . . "* R, 56.

3   *"Their rapidity of motion . . . "* R, 63.

3   *"[She] set right to work . . . "* R, 63-64.

3   *"I lay down to sleep . . . "* R, 66.

4   *"Whence I looked, down . . ."* R, 72.

5   *"No conception can be . . . "* Elisha Perkins, June 27, 1849: California National Historic Trail Sites—Chimney Rock (http://www.nps.gov/cali/cali/site2.htm).

5   *"standing edgewise . . . "* W. S. McBride, 1850: Martin 1985, part I:7.

6   *"a broken, rocky, mountainous . . . "* William Swain, July 6, 1849: H, 182.

6   *"It is difficult to account . . . "* A. J. McCall, June 29, 1849: Mc, 45.

6   *"elevated and notable . . . "* J. Goldsborough Bruff, August 1, 1849: RG, 60.

7   *"It had been so windy . . . "* Eliza Ann McAuley, July 4, 1852: Ho4, 60-61.

8   *"The word steep does not . . . "* E. W. Conyers, 1852: Oregon-California Trails Association Virtual Tour—Sublette Cutoff (http://www.octa-trails.org/JumpingOffToday/VirtualTour/SubletteCutoff.asp).

8   *"Of all countries for . . . "* Major Osbourne Cross, 1849: National Park Service—Experience the Oregon Trail (http://www.nps.gov/hafo/oregon/emguid1.htm).

8   *"In some places we . . . "* Ezra Meeker, 1852: National Park Service—Experience the Oregon Trail (http://www.nps.gov/hafo/oregon/emguid1.htm).

9   *"'The Oregon Trail' strikes . . . "* Wakeman Bryarly, July 16, 1849: Potter 1945, 157.

9   *"Our road this afternoon . . . "* John Hawkins Clark, August 5, 1852: B, 98.

10   *"Nothing but the hot . . . "* Bennett Clark, July 16, 1849: B, 131.

10   *"repay them for all . . . "* Margaret Frink, August 20, 1850: Ho2, 143-44.

11   *"From the top of the . . . "* Dan Carpenter, June 1850: John G. Mitchell, "The Way West," *National Geographic Magazine* (http://www.nationalgeographic.com/ngm/0009/feature2/zoom2.html).

11   *"This is the poorest . . . "* Dan Carpenter, August 1850: Mitchell 2000, 58.

CHAPTER 2: BETWEEN WINTER'S CHILL BRACKETS

My main sources for the history of the overland trails and the emigrants' trail experiences were McLynn 2002, Stewart 1962, and Unruh 1979. Brands 2002 was my main source of information on the life of John C. Frémont.

There exists a vast literature on the history of geologic thought and the ideas of Nicolaus Steno, James Hutton, Charles Lyell, and their contemporaries. My main sources were Carruthers 1999; Cutler 2003, 187–99; Gould 1987; Hutton 1788; Hutton 1795; Lyell 1830; McPhee 1998 (Book 1: Basin & Range, 66–99); and Repcheck 2003. See Cannon 1960 for an excellent summary of the uniformitarian-catastrophist debate.

Information on the emigrants' observations about geologic features and geologic processes is taken mostly from Martin 1985 (parts I–III), supplemented with my own reading of emigrant accounts.

13   *"The gold mania rages . . . "* New York Tribune, December 11, 1848: RG, xxxii.

14   *"I do not know where . . . "* H. Merrill, August 5, 1849: H, 199.

15   *"Love is hotter her[e] . . . "* John Lewis, 1852: Unruh 1979, 397.

15   *"The scenery through which . . . "* Harriet Ward, 1853: Levy 1990, 14.

16   *"The remembrance of scenery . . . "* Byron McKinstry, July 28, 1850: McK, 219–20.

16   *"I do not know when . . . "* A. J. McCall, June 15, 1849: Mc, 37.

16   *"To enjoy such a trip . . . "* anonymous emigrant, 1852: Unruh 1979, 414.

16   *"I would make a brave . . . "* Lavinia Porter, 1860, By Ox-team to California. Manuscript in the collection of the Bancroft Library, University of California, Berkeley, 55.

17   *"Oh, surely we are seeing . . . "* Lucy Cooke, 1852: Levy 1990, 16.

17   *"storm was decidedly . . . "* James Lyon, July 4, 1849: H, 158.

17   *"That desert is truly . . . "* Lucius Fairchild, 1849: FKL, 49.

17   *"[He] says he can't . . . "* John Edwin Banks, June 9, 1849: S, 17.

18   *"The fact is every . . . "* Richard May, August 14, 1848: B, 20.

18   *"It has been and . . . "* James Pritchard, July 24, 1849: M, 121.

23   *"This whole region of . . . "* Joseph Middleton, 1849: Martin 1985, part II:21.

23   *"All geologists know in . . . "* Stephen Jay Gould, 1987, 64.

25   *"There is presently laying . . . "* James Hutton, 1785, 50.

25   *"The strata formed at the . . . "* James Hutton, 1788, 263.

25   *"Time, which measures everything . . . "* James Hutton, 1788, 215.

25   *"no vestige of a beginning . . . "* James Hutton, 1788, 304.

26   *"The mind seemed to . . . "* John Playfair, 1803: Carruthers 1999, 86.

26   *"produce a degree of . . . "* John Playfair, 1805: Repcheck 2003, 205.

28   *"We know that one . . . "* Charles Lyell, 1830, vol. 1:80.

29   *"I always feel as if my . . . "* Charles Darwin, August 29, 1844 letter to Leonard Horner, reprinted in More Letters of Charles Darwin, 2 vols., edited by F. Darwin and A. C. Seward (London: John Murray, 1903), 2:117.

30   *"carries onward to the . . . "* Riley Root, 1848: Martin 1985, part III:28.

30   *"still in active operation . . . "* James Clyman, 1846: Martin 1985, part III:28.

30   *"in awe of Him . . . "* Martha Missouri Moore, July 23, 1860: Munkers 1989, 4.

30   *"torn by the rushing flood"* John Edwin Banks, June 12, 1849: S, 20.

30   *"A volcanic eruption must . . . "* Dan Gelwicks, 1849: Martin 1985, part II, 21.

30   *"thrown up by volcanic . . . "* Elizabeth Dixon Smith, July 7, 1847: Ho1, 123.

30   *"thrown up to a great . . . "* James Pritchard, June 8, 1849: M, 87.

30   *"their volcanic origin by . . . "* James Clyman, 1846: Martin 1985, part II, 21.

CHAPTER 3: ASCENDING THE PLAINS

The history of emigration along the Platte Valley is primarily from Mattes 1969. The geologic history of the assembly of the continental basement, including the Great Plains Orogeny and the Cheyenne Belt, comes from Baldridge 2004, Hoffman 1988, Houston 1993, Marshak 2000, McPhee 1998 (book 5, "Crossing the Craton"), and Snoke 1993.

31   *"We had good grass . . . "* Wakeman Bryarly, July 26, 1849: Potter 1945, 167.

32   *"The whole country around . . . "* Margaret Frink April 23, 1850: Ho2, 74–75.

33   *"[The] roads were thickly . . . "* Margaret Frink, May 20, 1850: Ho2, 85–86.

33   *"more islands half covered . . . "* Osbourne Cross, 1849: Martin 1985, part III:28.

33   *"were navigating two 'Mackinaw boats' . . . "* Edwin Bryant, June 11, 1846: Br, 83.

35   *"Still in camp, my husband . . . "* Amelia Stewart Knight, June 6, 1853: Schlissel 1982, 206–7.

35   *"Water poor white with clay . . . "* Amelia Hadley, May 22, 1851: Ho3, 63.

35   *"We encamped this afternoon . . . "* Edwin Bryant, June 15, 1846: Br, 93.

35   *"Our chief inconvenience here . . . "* Margaret Frink, May 24, 1850: Ho2, 88–89.

36   *"Wood is now very scarce . . . "* Tamsen Donner, June 16, 1846: NF, 21.

36   *"Burning with a lively . . . "* Edwin Bryant, June 9, 1846: Br, 80.

36   *"They emit a delicate . . . "* P. Pratt, July 1, 1849: H, 151.

36   *"It takes an average of . . . "* John King, June 16, 1850: H, pp. 150–51.

36   *"They are so great a . . . "* Joseph Warren Wood, May 23, 1849: Oregon Trail Education Resource Guide, published by the National Historic Oregon Trail Interpretive Center, Baker City, Oregon, 32.

36   *"The buffalo chips being . . . "* Edwin Bryant, June 22, 1846: Br, 103.

36   *"It is the duty of the . . . "* Wellman Packard and Greenberry Larison, 1850: H, 150.

36   *"We are beginning now . . . "* Edwin Bryant, June 4, 1846: Br, 72.

36   *"Experienced hunters aim . . . "* Edwin Bryant, June 18, 1846: Br, 95.

37   *"The casualties of buffalo . . . "* J. Goldsborough Bruff, June 26, 1849: RG, 24.

37   *"The sweetest and tenderest . . . "* William Swain, June 20, 1849: H, 155.

37   *"beef from a young . . . "* Edwin Bryant, June 18, 1846: Br, 96.

37   *"Not less than fifty . . . "* Lorenzo Sawyer, May 21, 1850: H, 153.

38   *"come out of their . . . "* Franklin Langworthy, May 28, 1850: H, 159.

38   *"The prairie dog is of . . . "* A. J. McCall, June 3, 1849: Mc, 28.

38   *"One of our hunters . . . "* A. J. McCall, June 17, 1849: Mc, 38.

38   *"The antelopes did not . . . "* Edwin Bryant, June 4, 1846: Br, 72.

38   *"You cannot slip up on . . . "* Charles Parke, May 23, 1849: Parke 1989, 18.

40   *"Dear wife, my heart bleeds . . . "* Joshua Sullivan, May 8, 1849: Rohrbough 1997, 52.

40   *"I miss you more than . . . "* Agnes Stewart, 1853: Schlissel 1982, 30.

40   *"O!! how I want . . . "* and *"William, if I could see . . . "* Sabrina Swain, August 24, 1849: H, 223.

CHAPTER 4: EXHUMED MOUNTAINS AND HUNGRY RIVERS

The geologic history of the Great Plains sedimentary layer cake and the Exhumation of the Rocky Mountains is from Maher, Englemann, and Schuster 2003, Diffendal 1987, and Swinehart and Loope 1987. The story of the Miocene rhinos killed by volcanic ash is from Mosel 2004.

A lot has been written on the debate over uplift versus climate change as the cause of the Exhumation of the Rocky Mountains. For reviews, see Epis and Chapin 1975 and Molnar and England 1990. For the paleobotanical argument for climate change as the cause of the Exhumation, see Wolfe, Forest, and Molnar 1998 and Gregory and Chase 1992. For the case for uplift based on basalt flow bubbles, see Sahagian, Proussevitch, and Carlson 2002. For the case for uplift based on the slope of the Cheyenne Tableland, see McMillan, Heller, and Wing 2006 and McMillan, Angevine, and Heller 2002.

51   *"fearful to look at . . . "* Margaret Frink, May 28, 1850: H02, 91.

54   *"When the wheels struck . . . "* A. J. McCall, June 5, 1849: Mc, 31.

54   *"one of the happiest . . . "* Margaret Frink, May 28, 1850: H02, 92.

56   *"We had a dreadful storm . . . "* Amelia Stewart Knight, May 17, 1853: Schlissel 1982, p. 205.

56   *"To day we had the . . . "* Elizabeth Dixon Smith, July 8, 1847: H01, 124.

56–57 *"the size of a* walnut . . . *," "writhing with the pain . . . ," "sundry bruised and gashed . . . ,"* and *" 'No great evil without . . . ' "* William Swain, June 20, 1849: H, 155–58.

58   *"The road hangs a little . . . "* Lewis Dougherty, 1848: RG, 593.

58   *"A general runaway and . . . "* Charles Scott, 1857: Oregon-California Trails Association Virtual Tour—Ash Hollow (http://www.octa-trails.org/JumpingOffToday/VirtualTour/AshHollow.asp).

58   *"We remained in camp . . . "* Margaret Frink, June 2, 1850: H02, 94.

58   *"Its width is not so great . . . "* Edwin Bryant, June 20, 1846: Br, 98–99.

59   *"No conception can be . . . "* Elisha Perkins, June 27, 1849, California National Historic Trail Sites—Chimney Rock (http://www.nps.gov/cali/cali/site2.htm).

59   *"line of pale and wintry . . . "* Edwin Bryant, June 22, 1846: Br, 103.

59   *"The soul must be cold . . . "* John Edwin Banks, June 12, 1849: S, 19.

60   *"One of these cliffs is . . . "* Zenas Leonard, 1839; *"Arrived at the Chimney . . .* " Nathaniel J. Wyeth, 1832; *"We are now in sight . . . "* William Marshall Anderson, 1834: all cited on an interpretive plaque at the Chimney Rock National Historic Site, Nebraska.

60   *"The geological processes . . . "* George Gibbs, 1849: Martin 1985, part III:29.

60   *"How came such an immense . . . "* Rufus Sage, 1841: Mitchell 2000, 43.

60   *"People say that it was . . . "* Rachel Larkin, 1853: P. Erickson, *Daily Life in a Covered Wagon* (New York: Puffin Books, 1994), 20.

61   *"is what remains of the . . . "* Edwin Bryant, June 21, 1849: Br, 101–2.

61   *"The whole country about . . . "* Joseph Stewart, 1849: Martin 1985, part III:28.

CHAPTER 5: BLACK HILLS AND BENT ROCK

Motives for westward migration in the days before the gold rush are discussed in McLynn 2002, 19-48, and in National Geographic Society 2005, 68-71.

The history of the horse and its significance for Plains Indians comes from West 1998, 49-57. The history of Indian-emigrant relationships and data on Indian-emigrant attacks comes from Unruh 1979, 156-200.

The geologic map of Love and Christianson 1985 is an essential starting point for learning Wyoming geology. Useful nonspecialist guidebooks to Wyoming geology include Blackstone 1988 and Lageson and Spearing 1991.

The theory of stream superposition traces back to John Wesley Powell (Powell 1875, 160-66).

The superposition history of Wyoming's rivers is discussed by Lageson and Spearing 1991, Mears 1993, Snoke 1993, and McPhee 1998 (Book 3: "Rising from the Plains").

Mears 1993 gives a thorough discussion of the origins of subsummit surfaces in Wyoming.

69   *"For what, then, do they . . . "* Horace Greeley, July 19, 1843: Unruh 1979, 38.

72   *"No man of information . . . "* *Missouri Republican,* June 1844: A. Hammond, "The Look of the Elephant," *Overland Journal* 23, no. 1 (2005), frontispiece.

72   *"the deep, rich, alluvial . . . "* Lansford Hastings, 1845: McLynn 2002, 27.

72   *"the fulfillment of our . . . "* John O'Sullivan, 1845: National Geographic Society 2005, 70.

73   *"What has miserable, inefficient . . . "* Walt Whitman, 1846: National Geographic Society, 71.

73   *"An American will build . . . "* Alexis de Tocqueville, *Democracy in America,* edited by J. P. Mayer, translated by George Lawrence (Garden City, NY: Doubleday, 1969), vol. 2, part 2, chap. 13, p. 536.

73   *"It is remarkable how . . . "* James Clyman, June 24, 1846: NF, 18.

73   *"We are now in sight . . . "* William Cornell, 1852: Moeller and Moeller 2001, 6.

73   *"Its snow covered top . . . "* Sarah Royce, 1849: R, 23.

74   *"becoming very hilly . . . "* William Swain, July 4, 1849: H, 169.

74   *"The general aspect of . . . "* Edwin Bryant, June 20, 1846: Br, 98.

74   *"The pretty young squaws . . . "* Andrew Griffith, June 3, 1850: H, 147.

74   *"Many of these women . . . "* Edwin Bryant, June 23, 1846: Br, 107-8.

75   *"When we camp at night . . . "* Sallie Hester, May 21, 1849: H01, 237.

75   *"You are now 640 miles . . . "* John M. Shively, 1846, *Guide to Oregon,* cited on an interpretive plaque at Fort Laramie National Historic Site, Wyoming.

76   *"Oh, what a treat it does . . . "* Lucy Cooke, 1852: Levy 1990, 10-11.

76   *"With the change in the . . . "* John C. Van Tramp, 1858, *Prairie and Rocky Mountain Adventures* (Columbus, OH: Segner & Condit, 1870), 351.

78   *"sixty miles over the worst . . . "* Sallie Hester, June 21, 1849: H01, 238.

78   *"Every day we pass good . . . "* Margaret Frink, June 10, 1850: H02, 98.

78   *"strewed with provisions which . . . "* A. J. McCall, June 21, 1849: Mc, 40.

78   *"Bar iron, black-smiths' anvils . . . "* A. J. McCall, June 23, 1849: Mc, 42.

78   *"Mighty glad were we . . . "*   William Swain, July 12, 1849: H, 185.

78   *"How I wish I was . . . "*   James Berry Brown, July 15, 1859: *Journal of a Journey Across the Plains in 1859,* edited by George A. Stewart (San Francisco: Book Club of California, 1970), 32.

82   *"It is difficult to . . . "*   A. J. McCall, June 29, 1849: Mc, 45.

83   *"had been rent by . . . "*   A. J. McCall, June 29, 1849: Mc, 45.

83   *"riven in two by . . . "*   Lucy Cooke, 1852: Levy 1990, 11.

83   *"there was fire below . . . "*   Charles Parke, June 29, 1849: Parke 1989, 44.

83   *"by volcanic force"*   Alonzo Delano, June 23, 1849: D, 99.

83   *"This is indeed wonderful . . . "*   Martha Missouri Moore, July 23, 1860: Munkers 1989, 4.

88   *"Men are daily drowned . . . "*   Franklin Langworthy, June 20, 1850: H, 188.

88   *"Some are making ferry . . . "*   Margaret Frink, June 16, 1850: Ho2, 100.

89   *"No sign of grass . . . "*   Henry Bloom, 1850: Natrona County Historical Preservation Commission 2001, 20.

90   *"collected this deposit . . . "*   A. J. McCall, June 26, 1849: Mc, 44.

90   *"We do not let our cattle . . . "*   J. R. Starr, July 1850: Moeller and Moeller 2001, 70.

90   *"seems to eat the lining . . . "*   Oliver Goldsmith, July 17, 1849: H, 189.

90   *"blind man might find . . . "*   Niles Searls, July 10, 1849: W, 127.

90   *"like the vertebrae of . . . "*   Sir Richard Burton, 1860: Richard Francis Burton, *The City of the Saints, and Across the Rocky Mountains to California* (New York: Harper & Brothers, 1862), 147.

90   *"immense piles of rocks . . . "*   Edwin Bryant, July 6, 1849: Br, 123.

91   *"We began to ascend it . . . "*   Sarah Royce, 1849: R, 25.

CHAPTER 6: TO THE BACKBONE OF THE CONTINENT

The age of Independence Rock and the Sweetwater Hills is from Love and Christiansen's 1985 geologic map.

The geologic story of the rise and fall of the Sweetwater Range, and the related boulder beds on top of Green and Crooks mountains, comes from Flanagan and Montagne 1993, Lageson and Spearing 1991, and Sayles 1983.

93   *"After breakfast, myself . . . "*   J. W. Nesmith, July 30, 1843: NF, 31-32.

102   *"It is grand, it is . . . "*   John Edwin Banks, June 29, 1849: S, 28.

102   *"a grand sight . . . "*   Lucy Cooke, 1852: Levy 1990, 11.

102   *"The road can be . . . "*   W. S. MacBride, June 9, 1850: H, 193-94.

103   *"A Mormon from the . . . "*   Joseph Middleton, July 28, 1849: H, 194.

103   *"well lined with carcasses . . . "*   Byron McKinstry, July 3, 1850: McK, 157.

104   *"We traveled late and . . . "*   William Swain, July 18, 1849: H, 189.

104   *"The wind, in addition . . . "*   Howard Stansbury, July 28, 1849: H, 189.

104   *"At dark, while I . . . "*   Margaret Frink, June 17, 1850: Ho2, 101.

105   *"We found a bank . . . "*   James Pritchard, June 18, 1849: M, 93.

105   *"clear and pure, and . . . "*   James Pritchard, June 16, 1849: M, 92.

105   *"so desirable a luxury . . . "*   J. Goldsborough Bruff, July 29, 1849: RG, 57.

105   *"We could hardly realize . . . "*   Margaret Frink, June 24, 1850: Ho2, 104-5.

105   *"I had looked forward . . . "*   Sarah Royce, August 4, 1849, R, 26-27.

106   *"There was a gloomy . . . "* A. J. McCall, July 5, 1849: Mc, 48.

106   *"got out the Star . . . "* Peter Decker, June 16, 1849: H, 199-200.

106   *"Music from a violin . . . "* Margaret Frink, June 24, 1850: Ho2, 105-6.

106   *"the most delicious ice cream . . . "* Charles Parke, July 4, 1849: Parke 1989, 46.

106   *"Just before sunset . . . "* Edwin Bryant, July 12, 1846: Br, 133-34.

107   *"cold, spiral and barren . . . "* Edwin Bryant, July 12, 1846: Br, 132.

CHAPTER 7: CORDILLERAN UPHEAVAL

The technical literature on the geologic evolution of the North American Cordillera is vast. My main sources for the discussion of the Farallon Plate and the Nevadan and Sevier orogenies were Baldridge 2004, DeCourten 2003, Dickinson 2001, Oldow et al. 1989, Snoke 1993, and Stanley 1999.

The history of Jurassic and Cretaceous terrane accretion in California is from Harden 2004 and Alt and Hyndman 2001.

The history of the Cretaceous Interior Seaway and the mid-Cretaceous superplume comes from Jahren 2002, Larson 1991, Larson 1995, and Steidtmann 1993.

The history of the Laramide Orogeny and the flat subduction theory comes from Brown 1993, Dickinson et al. 1988, English, Johnston, and Wang 2003, and Saleeby 2003.

The evidence for, and effects of, flat subduction beneath the Andes can be found in Fromm, Zandt, and Beck 2004 and Kay and Mpodozis 2001.

III   *"You may think you have . . . "* William Wilson, August 6, 1849: H, 204.

CHAPTER 8: MOST GODFORSAKEN COUNTRY

The information on Jim Bridger and the Green River Basin fur trappers comes from Alter 1962.

The geologic history of Eocene Lake Gosiute is taken from Ambrose 2004, Lageson and Spearing 1991, McDonald 1972, and McPhee 1998 (Book 3: "Rising from the Plains").

The geologic structure and evolution of the Overthrust Belt is based on Alt and Hyndman 1989, Lageson and Spearing 1991, Royce 1993, and Royce and Warner 1987.

137   *"a parting look at . . . "* Alonzo Delano, June 29, 1849: D, 116-17.

141   *"My companions denounce . . . "* A. J. McCall, July 9, 1849: Mc, 49.

142   *"scarcely possible to conceive . . . "* Edwin Bryant, July 15, 1846: Br, 139.

142   *"Is has been windy . . . "* A. J. McCall, July 8, 1849: Mc, 49.

143   *"a nasty, dirty place . . . "* Joseph Middleton, July 31, 1849: H, 208.

143   *"The desert over which . . . "* Alonzo Delano, July 1, 1849: D, 120.

145   *"Such a rough and . . . "* H. Merrill, July 14, 1849: H, 209.

145   *"The hill descending to . . . "* Franklin Starr, July 3, 1849: W. 157.

145   *"From the crest, down . . . "* J. Goldsborough Bruff, August 5, 1849: RG, 70-71.

145   *"rushed pell-mell down . . . "* William Swain, August 4, 1849: H, 209.

145   *"numerous dead oxen . . . "* J. Goldsborough Bruff, August 6, 1849: RG, 71.

145   *"high, deep, swift, blue . . . "* Margaret Frink, June 27, 1850: Ho2, 108.

145   *"While others are chasing . . . "* John Edwin Banks, July 11, 1849: p. 39.

145   *"A rope with pulleys . . . "* John B. Hill, 1850: Oregon-California Trails Association Virtual Tour—Sublette Cutoff (http://www.octa-trails.org/JumpingOff Today/VirtualTour/SubletteCutoff.asp).

146   *"Still in camp waiting . . . "* and *"The wagons are all . . . "* Amelia Stewart Knight, June 28–29, 1853: Schlissel 1982, 208.

146   *"an agreeable and picturesque . . . "* Edwin Bryant, July 15, 1846: Br, 139.

147   *"They had adopted the . . . "* Oliver Goldsmith, 1849: H, 210.

147   *"He was a born topographer . . . "* Grenville Dodge, December 11, 1904 (eulogy for Jim Bridger): Alter 1962, 341.

147   *"are generally well supplied . . . "* Jim Bridger, December 10, 1843 (letter to supplier): Alter 1962, 209.

147   *"two or three miserable . . . "* Edwin Bryant, July 17, 1846: Br, 142.

147   *"received me most cordially . . . "* A. J. McCall, July 14, 1849: Mc, 51.

147   *"sugar and coffee $1.00 . . . "* Richard Thomas Ackley, 1858: Oregon-California Trails Association Virtual Tour—Fort Bridger (http://www.octa-trails.org/JumpingOffToday/VirtualTour/FortBridger.asp).

148   *"elevated* buttes *of singular . . . "* and *"The plain appears at . . . "* Edwin Bryant, July 13, 1846: Br, 134.

151   *"The bluff is an immense . . . "* James Bennett, 1852: on Bureau of Land Management trailside interpretive plaque in the valley of Fontenelle Creek west of Rocky Gap.

152   *"We were delighted . . . "* Margaret Frink, July 4, 1850: Ho2, 111.

152   *"verdure of the country . . . "* William Swain, August 7, 1849: H, 211.

152   *"After having encamped . . . "* Israel Hale, July 8, 1849: W, 163.

152   *"We are entertained this . . . "* Augustus Burbank, July 4, 1849: H. 482.

153   *"Innumerable large black mice . . . "* J. Goldsborough Bruff, August 8, 1849: RG, 77.

154   *"The company had some . . . "* J. Goldsborough Bruff, August 10, 1849: RG, 85–86.

154   *"Dead animals all the . . . "* Byron McKinstry, July 19, 1850: McK, 194–95.

154   *"astonished how men ever . . . "* John Edwin Banks, July 20, 1849: S, 45.

154   *"compelled to let our . . . "* James Pritchard, June 24, 1849: M, 98.

154   *"so steep in places . . . "* Byron McKinstry, July 19, 1850: McK, 196.

CHAPTER 9: THE BEAR AND THE SNAKE

The geologic story of the Soda Springs area is taken from Alt and Hyndman 1989.

The geologic history of Yellowstone and its most recent "big three" eruptions comes from Lowenstern 2005, Lowenstern et al. 2005, and Smith and Siegel 2000.

The history of the Snake River Plain–to-Yellowstone caldera tract comes from Alt and Hyndman 1989 and Smith and Siegel 2000.

The mantle plume debate is a hot topic in geology right now, so my sources can only be considered a sample up to the time of this writing. Morgan 1971 gives Morgan's original mantle plume theory. My sources of information on the Yellowstone plume were Christiansen, Foulger, and Evans 2002, Humphreys et al. 2000, and Yuan and Dueker 2005. The story of the migrating Hawaiian hot spot is taken from Tarduno et al. 2003 and Steinburger, Sutherland, and O'Connell 2004. For general information on the mantle plume debate, both pro and con, see Anderson 2000, Dalton 2003, Foulger et al. 2005, Jones 2003, and Montelli et al. 2004.

155	*"This is the most . . . "* James Pritchard, June 25, 1849: M, 100.
155	*"The variety of scenery . . . "* Israel Hale, July 16, 1849: W, 172.
155	*"in wildfowl, ducks, geese . . . "* William Swain, August 9, 1849: H, 212.
159	*"All the wheels of . . . "* Margaret Frink, July 6, 1850: Ho2, 112.
159	*"only by having fastened . . . "* Cheyenne Dawson, 1841: McLynn 2002, 75.
159	*"boil up from the . . . "* Margaret Frink, July 9, 1850: Ho2, 114.
159	*"clear and sparkling and . . . "* James Pritchard, June 29, 1849: M, 102.
159	*"The water requires sugar . . . "* Franklin Starr, July 13, 1849: W, 174.
159	*"only needed lemon syrup . . . "* J. Goldsborough Bruff, August 17, 1849: RG, 91.
159	*"the same peculiar tingle . . . "* Byron McKinstry, July 24, 1850: McK, 205.
160	*"They foam and blubber . . . "* John Edwin Banks, July 23, 1849: S, 49.
160	*"will prove delicious and . . . "* Rufus B. Sage, 1842: Fanselow 2001, 129.
160	*"boiling forth . . . with a noise . . . "* Niles Searls, August 5, 1849: W, 175.
162	*"named for the resemblance . . . "* (fig. 9.4 caption) Bruff, August 17, 1849: RG, 91.
163	*"A large number of . . . "* J. Goldsborough Bruff, August 17, 1849: RG, 94.
164	*"They had lost about . . . "* George McCowen, July 15, 1854: B, 20.
164	*"a Scotchman, from Canada . . . "* J. Goldsborough Bruff, August 24, 1849: RG, 102.
165	*"were hospitably received . . . "* Margaret Frink, July 12, 1850: Ho2, 118.
165	*"We have now reached . . . "* Margaret Frink, July 12, 1850: Ho2, 118.
165	*"so broken and split . . . "* Margaret Frink, July 13, 1850: Ho2, 119.
165	*"Heat again became . . . "* Ezra Meeker, 1852: National Park Service—Experience the Oregon Trail (http://www.nps.gov/hafo/oregon/emguid1.htm).
165	*"one hundred and twenty . . . "* Margaret Frink, July 14, 1850: Ho2, 118-19.
166	*"It will fly so that . . . "* Elizabeth Dixon Smith, August 29, 1847: Ho1, 131.
166	*"more numerous than I . . . "* Israel Hale, July 24, 1849: W, 199.
166	*"as thick as flakes . . . "* Margaret Frink, July 11, 1850: Ho2, 117.
166	*"Our Fourth of July . . . "* James Pritchard, July 4, 1849: M, 108-9.
167	*"by turns, or all . . . "* Israel Lord, August 13, 1849: A. Hammond, "The Look of the Elephant," *Overland Journal* 20, no. 4 (2002), frontispiece.

CHAPTER 10: A BREAKING UP OF THE WORLD

Brock 2000, Curran 1982, and Hunt 1974 were my main sources of information for the history of emigration from the Snake River to the head of the Humboldt River.

The geologic story of City of Rocks National Reserve comes from Martin 1989 and Alt and Hyndman 1989.

For the story of the Basin and Range Orogeny and the formation of the Great Basin, I relied mostly on DeCourten 2003, with supplementary information from Colgan, Dimitru, and Miller 2004, Humphreys 1995, and Smith and Siegel 2000.

Seismic images of the Farallon Plate in the mantle below North America are given in Bunge and Grand 2000 and in Grand, van der Hilst, and Widiyantoro 1997.

Wolfe et al. 1997 present paleobotanical evidence for elevations in Nevada 16 million years ago being about one mile higher than today, before the crust stretched and subsided during the Basin and Range Orogeny.

Slemmons and Bell 1987 give the geologic story of the 1954 Fairview Peak and Stillwater Range earthquakes and their associated scarps.

Measurements of the ongoing stretching of the Basin and Range Province (the basis for fig. 10.9 and the associated text) are given in Flesch et al. 2000.

179   *"'The Oregon Trail' strikes . . . "*   Wakeman Bryarly, July 16, 1849: Potter 1945, 157.

179   *"If no bad luck happens . . . "*   Israel Hale, July 27, 1849: W, 202.

185   *"the least breeze or trampling . . . "*   James Bennett, August 11, 1850: H, 216-17.

185   *"How we do wish . . . "*   Helen Carpenter, 1857: Myres 1986, 30.

185   *"fresh water lobsters are . . . "*   Amos Batchelder, August 24, 1849: H, 219.

186   *"sublime, strange, and wonderful . . . "* (fig. 10.4 caption)   Margaret Frink, July 17, 1850: Ho2, 120-21.

186   *"You can imagine among . . . "*   Wakeman Bryarly, July 19, 1849, Potter 1945, 160-61.

186   *"a mass of common . . . "*   William Swain, August 26, 1849: H, 220-21.

188   *"fills the mind of man . . . "*   Leander Vaness Loomis, July 7, 1850: Oregon-California Trails Association Virtual Tour—City of Rocks (http://www .octa-trails.org/JumpingOffToday/VirtualTour/CityofRocks.asp).

189   *"the most rugged and . . . "*   Bernard Reid, August 12, 1849: B, 37.

189   *"would not have been . . . "*   Elijah Farnham, July 31, 1849: H, 231.

189   *"attached small trees to . . . "*   John Hawkins Clark, July 29, 1852: B, 38.

189   *"with long ropes held . . . "*   John Steele, August 9, 1850: B, 38.

190   *"How times change and . . . "*   A. J. McCall, August 7, 1849: Mc, 65.

190   *"Hardly a day passes . . . "*   Israel Hale, August 2-3, 1849: W, 208.

190   *"the effluvia of dead . . . "*   Franklin Langworthy, 1850: C, 58.

190   *"immense amount of property . . . "*   Lorenzo Sawyer, June 22, 1850: B, 39.

190   *"Had many applications . . . "*   William Rothwell, July 16, 1850: B, 42.

190   *"This evening a Boston . . . "*   Joseph Middleton, August 26, 1849: H, 221.

190   *"a stupendous amphitheater . . . "*   Franklin Langworthy, 1850: C, 58.

191   *"This valley certainly has . . . "*   William Gordon, July 23, 1850: B, 56.

191   *"This part of the country . . . "*   Mica Littleton, September 2, 1850: W, 210.

191   *"This morning we . . . came . . . "*   Margaret Frink, July 21, 1850: Ho2, 122-23.

192   *"Consequently it is generally . . . "* (fig. 10.5 caption)   Augustus Burbank, August 7, 1849: B, 50.

192   *"would be a glorious . . . "*   Byron McKinstry, August 9, 1850: McK, 247-48.

193   *"We are now encamped at . . . "*   John Hawkins Clark, August 2, 1852: B, 71.

193   *"It had evidently been . . . "*   Alonzo Delano, July 24, 1849: D, 152-53.

194   *"there had been a . . . "* (fig. 10. 6 caption)   Alonzo Delano, July 24, 1849: D, 152-53.

CHAPTER 11: MOST MISERABLE RIVER

Brock 2000, Curran 1982, and Hunt 1974 were my main sources for the history of emigration along the Humboldt River. Brock 2004 summarizes the history of the Applegate-Lassen Trail.

See Rieck 1995 for a comparative analysis of the topography and distance of the trails from the Humboldt River to the western side of the Sierra Nevada.

The geologic theories of stream superposition and stream antecedence trace back to John Wesley Powell (Powell 1875, 160-66).

205   *"Perhaps the Devil himself . . . "*   James Evans, 1850: FKL, 30.

205   *"Farewell to thee! . . . "*   Adison Crane, August 14, 1852: B, 198.

208   *"Meanest and muddiest . . . "*   Horace Belknap, 1850: B, 187.

209   *"We have seen more . . . "*   Leander Vaness Loomis, 1850, C, 49.

210   *"boiling hot & sent off . . . "* (fig. 11.3 caption)   Wakeman Bryarly, July 28, 1849: Potter 1945, 170.

210   *"God-forsaken, barren and . . . "*   A. J. McCall, August 22, 1849: Mc, 71.

211   *"The stream . . . begins to grow . . . "*   Lewis Beers, August 6, 1852: B, 114.

211   *"On whichever side . . . "*   Israel Lord, August 10, 1849: B, 96.

211   *"Our road this afternoon . . . "*   John Hawkins Clark, August 5, 1852: B, 98.

212   *"The scenery at this . . . "*   John Hawkins Clark, August 5, 1852: B, 100.

214   *"barely drinkable from saline . . . "*   Elisha Perkins, 1849: C. 41.

214   *"detestable; it is fairly . . . "*   Henry Sterling Bloom, 1850: C, 46.

214   *"We had not traveled . . . "*   Margaret Frink, August 10, 1850: Ho2, 131-32.

214   *"One can get an idea . . . "*   Heinrich Lienhard, 1846: C, 137.

214   *"so very crooked in . . . "*   Wakeman Bryarly, August 11, 1849: Potter 1945, 191.

215   *"Country all around to . . . "*   Henry Wellencamp, July 20, 1850: W, 222.

215   *"man will mire down . . . "*   John Birney Hill, August 6, 1850: W, 223.

215   *"Had very bad water . . . "*   William Gordon, August 6, 1850: B, 143.

215   *"Had to mow grass . . . "*   William Gordon, August 6, 1850: B, 143.

215   *"For about ten days . . . "*   Gilbert Cole, 1852: C, 46.

215   *"burying ground for horses . . . "*   Eleazer Stillman Ingalls, July 28, 1850: B, 172.

215   *"We frequently take our . . . "*   Franklin Langworthy, October 5, 1850: H, 242.

216   *"The scorching dry heat . . . "*   Edward Harrow, August 5, 1849: B, 159.

216   *"Through the day we . . . "*   John F. Riker, July 12, 1852: B, 118.

217   *"The eye tires & mind . . . "*   Bennett Clark, July 26, 1849: B, 131.

219   *"They were absolutely naked . . . "*   Reuben Cole Shaw, 1849: C, 35.

219   *"In the afternoon we met . . . "*   William G. Johnston, 1849: B, 81.

219   *"Roots, seeds, and grass . . . "*   John C. Frémont, August 29, 1843: Hunt 1974, 45.

220   *"These Diggers are a . . . "*   Israel Hale, August 15, 1849: W, 226.

220   *"We fell in company with . . . "*   Alonzo Delano, August 6, 1849: D, 168.

220   *"The emigrant had been . . . "*   Franklin Langworthy, September 24, 1850: B, 138.

220   *"Ordered the men to shoot . . . "*   Israel Lord, September 2, 1849: B, 112.

221   *"I see at least a dozen . . . "*   Byron McKinstry, August 15, 1850: McK, 259.

221   *"We have been no little . . . "*   Thomas Christy, July 16, 1850: B, 156.

221   *"Often, almost daily, some . . . "*   Eleazer Stillman Ingalls, 1850: C, 40–41.

221   *"I presume that twenty . . . "*   Byron McKinstry, August 18, 1850: McK, 264.

221   *"noticed several dead horses . . . "*   Eleazer Stillman Ingalls, 1850: C, 40–41.

221   *"We came on the bank . . . "*   Lemuel C. McKeeby, 1850: C, 41.

221   *"Here, on the Humboldt . . . "*   Horace Greeley, 1860, 272.

222   *"The whole environment as . . . "* (fig. 11.7 caption)   Oliver Goldsmith, September 22, 1849: H, 253.

224   *"Our great want, now . . . "*   John Hawkins Clark, August 13, 1852: B, 170.

224   *"It would almost seem . . . "*   Lorenzo Sawyer, 1850: C, 129.

225   *"The grass . . . cured rapidly . . . "*   Margaret Frink, August 13, 1850: Ho2, 135.

225   *"a vast Quagmire or . . . "*   James Pritchard, July 25, 1849: M, 122.

225   *"a mud lake ten miles . . . "*   Reuben Cole Shaw, August 26, 1849: B, 195.

225   *"the end of the most . . . "*   Margaret Frink, August 14, 1850: Ho2, 136.

CHAPTER 12: THE WORST DESERT YOU EVER SAW

Brock 2000, Curran 1982, Fey, King, and Lepisto 2002, and King 2003 were my main sources for the emigrants' experience of the Forty-Mile Desert and ascent of the Sierra Nevada.

The geologic history of Pleistocene Lake Lahontan comes from Adams and Wesnousky 1998, DeCourten 2003, and Rehels 1999.

See Sharp and Glazner 1997b for discussion of tufa deposits associated with Great Basin pluvial lakes.

The evidence for recent (last 5 million years) uplift of the Sierra Nevada range is from Harden 2004, 159–63, with supplementary information from Stock, Anderson, and Finkel 2004 and Wakabayashi and Stock 2003. However, Mulch, Graham, and Chamberlain 2006 argue that the Sierra Nevada existed as high topography as early as 50 million years ago, and Crowley, Koch, and Davis 2004 present isotopic evidence that the Sierra Nevada was high enough by 18 million years ago to trap significant rainfall.

Information about the Eastern Sierra Frontal Fault System, including the Owens Valley 1872 quake and the Genoa Fault, comes from DeCourten 2003, Kent et al. 2005, Orndorff, Wieder, and Flikorn 2001, Sharp and Glazner 1997a, and Unruh, Humphrey, and Barron 2003.

The history of the Sierra Nevada–Great Valley block and its travels from Las Vegas to California (fig. 12.10) is based on Wernicke and Snow 1998.

See Boyd, Jones, and Sheehan 2004, Saleeby and Foster 2004, and Zandt et al. 2004 for discussions of how mantle processes and lithospheric delamination below the Sierra Nevada appear to have driven recent uplift of the range.

229   *"The desert! You must . . . "*   Eleazer Stillman Ingalls, August 5, 1850: FKL, 120.

229   *"From one extremity . . . "*   Mark Twain, *Roughing It* (New York: American Publishing Co., 1872), 121.

232   *"would not be as bad . . . "*   J. S. Shepherd, 1850: C, 143.

232   *"All are preparing and . . . "*   John Wood, 1850: FKL, 33.

232   *"strong with salt and . . . "*   Margaret Frink, August 14, 1850: Ho2, 136.

232   *"These wells although . . . "*   Augustus Burbank, August 31, 1849: B, 198.

233   *"Dead horses and oxen . . . "*   John T. Clapp, July 15, 1850: B, 197.

234   *"Every thing around is . . . "* Edwin Bryant, August 19, 1846: Br, 217.

234   *"The rocks had a peculiar . . . "* Heinrich Lienhard, 1846: C, 145.

237   *"One of the most disagreeable . . . "* Alonzo Delano, August 4, 1849: D, 166.

237   *"where the ground shoots . . . "* Jo Utter, 1847: C, 145.

237   *"there are a great many . . . "* Franklin Starr, August 23, 1849: W, 255.

238   *"lined with dead cattle . . . "* Lucius Fairchild, 1849: FKL, 49.

239   *"The sun had risen and . . . "* William Woodhams, July 10, 1854: A. Hammond, "The Look of the Elephant," *Overland Journal* 22, no. 1 (2004), frontispiece.

239   *"waiting for death . . . "* John Edwin Banks, September 2, 1849: S, 78.

239   *"All our traveling experience . . . "* Bennett Clark, 1849: FKL, 48.

239   *"If ever I saw heaven . . . "* Lydia Waters, 1855: King 2003, 133.

239   *"It is wonderful to . . . "* John Edwin Banks, September 2, 1849: S, 78.

239   *"No one can imagine . . . "* Elisha Perkins, 1849: C, 152.

240   *"intensely brackish, bitter . . . "* William G. Johnson, 1849: FKL, 121.

241   *"The sand hills are reached . . . "* Eleazer Stillman Ingalls, August 5, 1850: FKL, 120.

242   *"filled the entire roadside . . . "* John Hawkins Clark, 1852: C, 181.

242   *"The destruction of property . . . "* Franklin Langworthy, 1850: C, 181.

242   *"For many weeks we . . . "* Margaret Frink, August 16, 1850: Ho2, 138–39.

242   *"The day was oppressively . . . "* James Pritchard, July 27, 1849: M, 125.

242   *"howling for water . . . "* William Kelly, 1849: C, 184.

242   *"After the nauseous stuff . . . "* Margaret Frink, August 16, 1850: Ho2, 139.

243   *"rush up, half crazed . . . "* Jasper Hixson, 1849: C, 185.

243   *"Before we got to . . . "* James Carpenter, 1852: C, 184.

243   *"all to strap the . . . "* Thomas Christy, 1850: FKL, 149.

243   *"At Ragtown, to our . . . "* Franklin Langworthy, 1850: FKL, 124.

243   *"Its water was clear . . . "* Margaret Frink, August 17, 1850: Ho2, 140.

243   *"God of Heaven! . . . "* John Wood, 1850: C, 183.

243   *"the awfulest country . . . "* Thomas Christy, 1850: FKL, 115.

243   *"From the summit of . . . "* William Kelly, 1849: Oregon-California Trails Association Virtual Tour—Mormon Station (http://www.octa-trails.org/Jumping OffToday/VirtualTour/MormonStation.asp).

249   *"was exactly like marching . . . "* Elisha Perkins, 1849: C, 170.

249   *"one of the grandest . . . "* James Pritchard, August 3, 1849: M, 130.

250   *"We never tire of . . . "* Margaret Frink, August 21, 1850: Ho2, 144.

250   *"Trees once more! . . . "* Andrew Grayson, 1846: FKL, 88.

CHAPTER 13: INTO THE LAND OF GOLD

The story of the Donner Party's entrapment is from Fey, King, and Lepisto 2002, 89–98, and McLynn 2002, 326–70.

Curran 1982 and Fey, King, and Lepisto 2002 were my main sources for the story of ascending the Sierra Nevada on the Truckee and Carson Routes, with supplemental information from Graydon 1986 (for the Truckee Route) and Tortorich 2002 (for the Carson Route).

The geologic story of California's gold comes mainly from Alt and Hyndman 2000, Harden 2004, and Hill 2002. Silva 1986 gives a comprehensive treatment of placer mining. The information about the auriferous gravels and hydraulic mining is from Hill

2002, McPhee 1998 (Book 4: "Assembling California"), and Wyckoff 1999. For general information on the hydrothermal origin of gold-bearing quartz veins, see Kirkemo, Newman, and Ashley 2006 and Muntean 2006. Data on California gold production and numbers of miners are from Holliday 1999, 151-53.

The role of California gold in the American Civil War is discussed by Brands 2002, 394-405.

251   *"I wish California had . . . "*   James Wilkins, September 9, 1849: NF, 50.

251   *"We arrived at the . . . "*   Sallie Hester, September 14, 1849: H01. 242.

251   *"stumps from ten to . . . "*   A. J. McCall, September 7, 1849: Mc, 79.

253   *"found many human bones . . . "*   Wakeman Bryarly, August 21, 1849: Potter 1945, 202.

253   *"Cannibal Cabins"*   a phrase used by A. J. McCall, September 7, 1849 (Mc, 79), John Edwin Banks, September 14, 1849 (S, 87), and Charles Long, 1849 (FKL, 106).

253   *"Standing at the bottom . . . "*   Edwin Bryant, August 26, 1846: Br, 230.

256   *"We came to a rim . . . "*   Benjamin Bonney, 1846: FKL, 101.

257   *"When we came to . . . "*   David Hudson, 1845: FKL, 102.

257   *"as steep as the roof . . . "*   Joseph Hackney, 1849: FKL, 109.

258   *"The aspect of the country . . . "*   John Edwin Banks, September 14, 1849: S, 87.

258   *"We felt a real relief . . . "*   John Steele, 1850: C, 174.

258   *"As I stood there . . . "*   A. J. McCall, September 7, 1849: Mc, 80-81.

259   *"most decidedly the worst . . . "*   George Willis Read, 1850: C, 194.

259   *"mules and wagons staggering . . . "*   William Kelly, 1849: C, 194.

259   *"It made one's flesh . . . "*   William Kelly, 1849, Dodd and Gnass 1996, 53.

259   *"the wildest hallooing . . . "*   William Johnston, 1849, Dodd and Gnass 1996, 53.

262   *"the most dreaded by . . . "*   Franklin Langworthy, 1850: C, 195.

262   *"Had we met such . . . "*   William Kelly, 1849: C, 196-97.

262   *"The road is crooked . . . "*   Franklin Langworthy, 1850: Dodd and Gnass 1996, p. 55.

262   *"the Horses [were] unable to . . . "*   J. Wesley Jones, 1851: FKL, 167.

263   *"Nothing in nature I . . . "*   James Pritchard, August 7, 1849: M, 135.

263   *"The scenery is sublime . . . "*   James Wilkins, September 26-28, 1849: Dodd and Gnass 1996, 57.

263   *"The Summit is crossed . . . "*   Niles Searls, October 1, 1849: W, 272.

263   *"We had expected an . . . "*   Margaret Frink, August 31, 1850: H02, 153.

264   *"We here saw for the . . . "*   Joseph Hackney, September 11, 1849: NF, 55.

264   *"Wherever we turned, we . . . "*   Alonzo Delano, 1849: D, 281-82.

264   *"The price of provisions . . . "*   Tom Archer, *Recollections of a Rambling Life*, 1897: Brands 2002, 211.

264   *"Even in the most . . . "*   Alonzo Delano, 1849: D, 373.

265   *"Oh Caroline, I can't . . . "*   Andrew Orvis, September 15, 1850: H, 350.

265   *"Say to all my friends . . . "*   Jerome Dutton, December 28, 1850: H, 350.

265   *"I really hope that no . . . "*   Dr. Isaac Lord, August 11, 1850: H, 352-53.

269   *"Miners are in the . . . "*   Alonzo Delano, 1849: D, 373.

271   *"The best veins abound . . . "*   Alonzo Delano, 1849: D, 377.

274   *"Quartz and granite appear . . . "*   Alonzo Delano, 1849: D, 378.

CHAPTER 14: CONTINGENT HISTORY

276   *"Contingency is central to . . . "* David Hackett Fisher, *Paul Revere's Ride* (Oxford: Oxford University Press, 1994), xv.

278   *"been compelled to yield . . . "* Robert E. Lee, "Farewell Address to Army of Northern Virginia, Appomattox Court House," April 9, 1865.

278   *"It is a question whether . . . "* John Bidwell, 1890: *Addresses, reminiscences, etc. of General John Bidwell.* Complied by Charles C. Royce, 1906. Reproduced in *California as I Saw It: First-Person Narratives of California's Early Years, 1849–1900.* Library of Congress General Collections, digital call number: calbk 046.

EPILOGUE

Biographical information on the individual emigrants comes from the following sources. Alonzo Delano: Sierra Nevada Virtual Museum of Sierra Community College, Rocklin, California (http://www.sierranevadavirtualmuseum.com/docs/specialex/biographies/delanoa.htm); Margaret Frink: Holmes 1983, vol. 2, 55–58, 159–69; William Swain: Holliday 1981, 54–60, 409–50; Byron McKinstry: McKinstry 1975, 23–58, 381–85; Anselm J. McCall: McCall 1882, 3–8, and McCall 1883, 1–46; J. Goldsborough Bruff: Read and Gaines 1949, xix–lxxii; Edwin Bryant: Bryant 1848, v–xx (Thomas D. Clark's introduction to the 1985 University of Nebraska Press republication of Bryant's 1848 book).

279   *"delving among the mountains . . . "* Alonzo Delano, 1849: D, 380.

279   *"a great attraction to men . . . "* Margaret Frink, November 30, 1850: Ho2, 164–65.

279   *"The progress of time . . . "* Margaret Frink, undated: Ho2, 167.

279   *"I am coming back . . . "* William Swain, April 11, 1849: H, 63.

280   *"no faith in the . . . "* A. J. McCall, 1882: Mc, 3.

280   *"The last six weeks . . . "* A. J. McCall, October 30, 1849: McCall 1883, 27.

281   *"fairly reveled in gold . . . "* Mark Twain, *Roughing It* (1871), chap. 57.

281   *"My mouth fairly watered . . . "* J. Goldsborough Bruff, April 8, 1850: RG, 339.

281   *"I had 'seen the elephant' . . . "* J. Goldsborough Bruff, July 20, 1851: RG, 523.

GLOSSARY OF
KEY GEOLOGIC TERMS

accretion (of terranes). The process, usually associated with subduction, of adding terranes to one another or to the edge of a continent. Terrane accretion causes continents to grow larger at their edges and is responsible for assembling much of western North America, including California.

andesite. An igneous rock formed from solidified lava, intermediate in mineral and chemical composition between basalt and rhyolite, and dominated by small crystals of plagioclase feldspar and amphibole. The same magma forms a rock called diorite if it cools slowly underground rather than erupting.

ash (volcanic). Tiny particles, usually with a glassy texture, ejected from volcanoes during violent eruptions. Formed as escaping gases blast apart the magma into a fine particulate spray.

auriferous gravels. Gold-bearing gravels exposed throughout much of the western Sierra Nevada, representing the channels of large rivers that flowed west from Nevada to the Pacific Ocean during Eocene time, before the uplift of the Sierra Nevada range.

basalt. A very common black volcanic rock formed from solidified lava and composed of sand-sized crystals of mostly plagioclase feldspar, pyroxene, and olivine. The same magma that forms basalt becomes a rock called gabbro if it cools slowly underground rather than erupting.

Basin and Range Orogeny. The mountain-building episode that created the Basin and Range Province by east-west-directed stretching of the crust, beginning about 20 million years ago (early in Miocene time) and continuing today.

Basin and Range Province. A region of north-south-oriented, fault-bounded mountains separated by sediment-filled valleys (basins), encompassing all of Nevada and portions of Wyoming, Idaho, Oregon, Utah, Arizona, New Mexico, California, and northern Mexico.

basement. The oldest rocks in a continental area, generally underlying younger volcanic and/or sedimentary formations except where pushed up and exposed in mountains.

Consisting of assorted metamorphic and igneous rocks that appear to have been assembled, in part, by tectonic collisions between ancient volcanic island arcs and other crustal fragments.

basement-cored uplift. A mountain formed of basement rock that has been pushed upward, usually by sideways compression along large thrust faults. The Rocky Mountain Foreland Ranges of the western United States and the Pampean Ranges of Argentina are classic examples.

batholith. An igneous mass of many fused plutons with an exposed area of at least 40 square miles. Generally long and narrow, and often composed of granite. Batholiths form deep underground and are thus exposed only where overlying rocks have eroded away due to mountain uplift (example: Sierra Nevada batholith).

caldera. A large, circular depression or crater, often miles across, formed by a violent volcanic eruption followed by collapse of the ground surface.

cirque. The head of a glacial valley, usually with a curving, amphitheater-like shape and having steep slopes along the upper edges and a flat or hollowed-out base often occupied by a lake.

crust. The outermost layer of the Earth, consisting of either continental crust (20 to 40 miles thick and mostly of granitic composition) or oceanic crust (4 to 7 miles thick and mostly of basaltic composition).

diorite. An igneous rock intermediate in mineral and chemical composition between granite and gabbro, dominated by plagioclase feldspar and amphibole, and formed by magma cooling slowly underground. The same magma forms a rock called andesite if it erupts as lava onto the Earth's surface.

Exhumation of the Rocky Mountains. The uncovering, by river and wind erosion, of the once deeply buried Rocky Mountains to create the rugged landscape that we see today, where rivers commonly slash through mountain ridges (*see* stream superposition; subsummit surface).

Farallon Plate. A large oceanic plate that once filled much of the Pacific Ocean basin. As Pangaea broke up and North America migrated west, the Farallon Plate subducted underneath the continent's western edge to create the North American Cordillera.

fault. A planar or gently curved fracture in the Earth's crust where the rocks on either side have shifted measurably. Energy released by shifting along faults produces earthquakes.

fault scarp. A cliff or ridge formed by movement along a fault during an earthquake. Represents the exposed surface of a fault.

fold-and-thrust belt. The tectonic zone of a mountain belt between the volcanic arc and the foreland basin, characterized by intense sideways compression, where rock strata have been folded and pushed toward the continental interior along large thrust faults.

foreland basin. An elongate depositional basin that lies between a mountain belt and the continental interior. Generally formed by the weight of the mountain belt bowing down the crust to create a depression that fills with rock debris eroded from the mountain belt.

gabbro. A black or black-green igneous rock composed of large crystals of mostly plagioclase feldspar, pyroxene, and olivine, formed where magma cools slowly underground. The same magma that forms gabbro makes basalt if it erupts onto the Earth's surface as lava.

geyser. A fountain of hot water and steam ejected periodically from below the Earth's surface. The heat generally results from contact of groundwater with hot rock.

gneiss. A metamorphic rock with a distinctive banded appearance formed by alternating layers of light-colored and dark-colored minerals. Common in the deep cores of mountain belts and in the continental basement.

granite. Broadly defined, granite is a light-colored igneous rock composed mostly of large crystals of quartz, pinkish orthoclase feldspar, white or gray plagioclase feldspar, and muscovite mica, along with a scattering of darker crystals of biotite mica and amphibole. Technically, true granite is rare and much "granite" is actually granodiorite. But the term *granite* is so widespread that a broader definition is coming into favor. Granite (broadly defined) forms miles underground from the solidification of silica-rich magma and is the main rock of plutons and batholiths.

Great Plains Orogeny. A mountain-building episode from roughly from 1.8 to 1.6 billion years ago associated with the serial collision of volcanic island arcs to assemble the continental basement below the Great Plains. Before the Great Plains Orogeny, most of the North American continent south and east of Wyoming did not exist.

greenstone. A rock, often originally basalt or gabbro, altered by metamorphism so that greenish minerals (particularly chlorite) have grown in it to impart a greenish color. Greenstone is common in the North American basement and often forms elongate belts that may represent pieces of ancient ocean floor caught up between colliding volcanic island arcs.

hot spot. A point-source of concentrated volcanic activity that appears to stay in one place as a tectonic plate moves across it, leading to age-progressive lines of volcanoes (such as the Hawaiian-Emperor chain) or calderas (such as the Snake River Plain–to–Yellowstone caldera tract). Hot spots are widely thought to be located above mantle plumes (*see* mantle plume).

hydrothermal. Any process or activity involving high-temperature groundwater, especially the alteration and emplacement of minerals and the formation of hot springs and geysers (*see* vein).

Ice Age. An informal term for the glaciated times of the Pleistocene Epoch (1.8 million to 10,000 years ago), when the Earth was periodically cooler than usual and large areas of the continents were covered by ice.

igneous rock. A rock formed by solidification of magma either underground or on the Earth's surface.

isostatic adjustment. The mechanism whereby areas of the Earth's crust "float" in a buoyant state in the denser rock of the underlying mantle, rising up or sinking down to achieve equilibrium.

Laramide Orogeny. A Late Cretaceous to Eocene age mountain-building episode that squeezed blocks of basement rock upward to form the Rocky Mountain Foreland Ranges of Montana, Wyoming, Colorado, and New Mexico.

lode. A mining term referring to a vein or a zone of veins (*see* vein).

magma. Molten rock material formed within the Earth that becomes igneous rock upon cooling and solidification. Magma that erupts onto the surface is called lava.

mantle. The 1,800-mile-thick region between the crust and the core, forming roughly 80 percent of the volume of the Earth. The mantle is mostly solid rock, but so hot that the rock rises and sinks in slow convective cycles.

mantle plume. A narrow, columnar mass of hotter-than-normal mantle rock that rises through the mantle to form a point-source of intense volcanic activity at the Earth's surface. Mantle plumes are theoretical entities. Their existence is implied by hot spots, but they have yet to be widely detected (*see* hot spot).

metamorphic rock. A rock formed by the alteration, in a solid state, of a preexisting rock by heat and/or pressure and/or fluid interactions deep underground.

mid-ocean ridges. Broad and continuous ridges on the floors of all major ocean basins, 300 to 3,000 miles wide, with rift valleys running down the centers. The seafloor spreads from the rift valleys, forming new oceanic crust through the eruption of basalt lava (*see* seafloor spreading).

Nevadan Orogeny. A Jurassic age mountain-building episode in western North America triggered by the continent's westward migration out of Pangaea and the initiation of Farallon Plate subduction along the west coast.

oceanic plateau. An extensive region of ocean floor made up of particularly thick accumulations of pillow basalt (up to 20 miles thick).

oceanic trench. A deep, linear depression on the seafloor formed by subduction, where an oceanic plate bends down underneath another plate and plunges into the Earth's interior (*see* subduction zone).

ophiolite. Rock of the ocean floor that has been transported up onto a continent, usually in areas of subduction or continental collision. May include (in original vertical order from top to bottom): deep-sea sediments, pillow basalt, sheeted dikes, gabbro from the lower reaches of the oceanic crust, and/or peridotite from the uppermost mantle. The basalt and gabbro may be altered to serpentinite (*see* serpentinite).

orogenesis. The tectonic processes that collectively result in the formation of mountain belts, typically involving faulting, folding, metamorphism, and magma generation on a large scale.

Pangaea. The supercontinent that began breaking up about 200 million years ago, leading to the opening of the Atlantic Ocean and the present distribution of the continents.

peridotite. An igneous rock composed of mostly of olivine with smaller amounts of pyroxene and amphibole and little or no feldspar. Thought to be the main rock of the upper mantle.

pillow basalt. Bulbous, pillow-shaped lava formations generated where basaltic lava erupts underwater. Most of the ocean floor beneath a veneer of younger sediment layers is paved with pillow basalt originally formed by seafloor spreading at mid-ocean ridges.

placer. A deposit formed where heavy mineral grains are concentrated by the agitating motion of stream flow or waves. Placers are sources of gold, diamonds, platinum, and other valuable minerals.

plate tectonics. The well-established theory that holds that the Earth's outer rocky shell is broken up into several dozen individual plates, 50 to 100 miles thick, that move and interact to produce earthquakes, volcanoes, and most of the major geographic features of the planet, including mountain belts.

pluton. A mass of igneous rock, often bulbous in shape and several miles across, formed where magma cools and solidifies deep underground. Plutons are exposed only where overlying rocks have eroded away, often by the uplift of mountains. Most plutons have a composition toward the granitic end of the igneous rock spectrum.

rhyolite. A volcanic rock, light in color and rich in silica, composed of small mineral grains or volcanic ash (example: rhyolite tuff). The same magma forms granite when it cools and solidifies deep underground rather than erupting.

schist. A common metamorphic rock (that is, a rock changed from a preexisting state by heat and pressure), characterized by distinctive layering produced through the parallel arrangement of flat minerals, particularly micas.

seafloor spreading. The mechanism by which new seafloor crust is created at mid-ocean ridges, where two plates diverge and magma wells up into the gap. Measured spreading rates range from a fraction of an inch to four inches per year and may have been faster earlier in the Earth's history.

sedimentary rock. A rock formed from weathered and eroded pieces of preexisting rocks that have been transported by wind, water, or ice, and deposited and cemented together. (Sandstone and conglomerate are examples.) Also includes rocks chemically precipitated out of water. (Limestone and gypsum are examples.)

serpentinite. A rock composed mostly of serpentine-group minerals, which in turn form largely by hydrothermal alteration of minerals in basalt and gabbro—the main rocks of the oceanic crust. Serpentine minerals are commonly greenish and soft, with a greasy or silky luster and a smooth, soapy feel (*see* ophiolite).

Sevier Orogeny. A mountain-building episode in the western United States, primarily of Cretaceous age, that took place after the Nevadan Orogeny to the west but before the Laramide Orogeny to the east. Characterized by massive eastward-directed folding and thrusting of rock. The eastward limit of the Sevier Orogeny is marked by the Overthrust Belt of Utah, Idaho, and Wyoming.

sink. A bowl-like valley between mountains in a desert region where a river ends. Sinks may contain saline lakes during times of high river flow but are usually dry and coated with mud and salt during times of low or absent flow. Common in the Great Basin.

strata. Layers of sedimentary rock, originally laid down horizontally, but which may be faulted, tilted, and folded during orogenesis.

stream superposition. A process by which streams establish themselves on smooth sedimentary layers that bury preexisting ridges and then cut down through the ridges as the region erodes (*see* Exhumation of the Rocky Mountains).

subduction zone. The region where an oceanic plate bends down beneath an adjacent plate and heads into the Earth's interior, forming an oceanic trench. Characterized by frequent earthquakes and volcanism in the adjacent volcanic arc (*see* volcanic arc).

subsummit surface. An eroded bench or step on the mountainsides of the Wyoming Rockies at elevations of 9,000 to 11,000 feet, often interpreted to reflect the maximum burial level of the Rockies prior to the Exhumation (*see* Exhumation of the Rocky Mountains).

suture. A zone of faults and fractures along which two terranes have collided and joined (*see* terrane).

tectonics. The study of the processes and forces that cause movement and deformation of the Earth's crust on a large scale (*see* plate tectonics).

terrane. A block of the Earth's crust, surrounded by faults, whose geologic history is distinct from adjacent crustal blocks, often because it has traveled from a distant area. Continents grow by the accretion of terranes at their edges.

thrust fault. A fault in which the rock above the fault moves up and over the rock below, usually because of sideways compression of the Earth's crust.

travertine. A form of limestone (calcium carbonate) precipitated from dripping water in caves and around hot springs where cooling, carbonate-saturated groundwater emerges onto the surface. Similar to tufa but less porous.

trona. A hydrated sodium bicarbonate mineral, fibrous, white to light brown in color, formed from evaporation of lake water. A major commercial source of sodium carbonate compounds.

tufa. A form of limestone (calcium carbonate) forming porous encrustations around the mouths of hot or cold seeps or springs, particularly underneath lakes or along lakeshores. Similar to travertine but softer and more porous.

tuff. A rock composed of pyroclastic particles; that is, particulate material blown into the air during a volcanic eruption (*see* ash).

uniformitarianism. The concept that the processes that have shaped the Earth in the geologic past are essentially the same as those operating today.

vein. A mineral filling in a rock fracture or fault, often of hydrothermal origin.

volcanic arc. A line of active volcanoes that forms parallel to an oceanic trench where subduction is taking place. Volcanic arcs form on the edges of continents wherever oceanic plates subduct beneath the continents (example: Andes). They also form in the ocean wherever an oceanic plate subducts beneath another oceanic plate (example: Aleutian Islands, Mariana Islands), in which case they are called volcanic island arcs.

BIBLIOGRAPHY

EMIGRANT HISTORY

Alter, J. C. 1962. *Jim Bridger*. Norman: University of Oklahoma Press.

Brands, H. W. 2002. *The Age of Gold: The California Gold Rush and the New American Dream*. New York: Doubleday.

Brock, R. K., ed. 2000. *Emigrant Trails West: A Guide to the California Trail from the Raft River to the Humboldt Sink*. Reno, NV: Trails West.

———, ed. 2004. *Emigrant Trails West: A Guide to the Applegate Trail, the Southern Road to Oregon*. Reno, NV: Trails West.

Brown, R. 2004. *Historic Inscriptions on Western Emigrant Trails*. Independence, MO: Oregon-California Trails Association.

Bryant, E. 1848. *What I Saw in California*. D. Appleton & Co. Reprint, Lincoln: University of Nebraska Press, 1985.

Curran, H. 1982. *Fearful Crossing: The Central Overland Trail through Nevada*. Las Vegas: Nevada Publications.

Delano, A. 1854. *Life on the Plains and among the Diggings*. Auburn, NY: Miller, Orton & Mulligan. Reprint, Alexandria, VA: Time-Life Books, 1981.

Dodd, C. H., and J. Gnass. 1996. *California Trail: Voyage of Discovery*. Las Vegas: KC Publications.

Fanselow, J. 2001. *Traveling the Oregon Trail*. 2nd ed. Guilford, CT: Globe Pequot Press.

Faragher, J. M. 1979. *Women and Men on the Overland Trail*. New Haven, CT: Yale University Press.

Fey, M., R. J. King, and J. Lepisto. 2002. *Emigrant Shadows: A History and Guide to the California Trail*. Virginia City, NV: Western Trails Research Association.

Franzwa, G. M. 1988. *The Oregon Trail Revisited*. 4th ed. Tucson, AZ: Patrice Press.

———. 1999. *Maps of the California Trail*. Tucson, AZ: Patrice Press.

Graydon, C. K. 1986. "Trail of the First Wagons Over the High Sierra." *Overland Journal* 4, no. 2:4–15.

Greeley, Horace. 1860. *An Overland Journey from New York to San Francisco in the Summer of 1859*. New York: C.M. Saxton, Barker & Co. Reprint, University of Nebraska Press, 1999.

Holliday, J. S. 1981. *The World Rushed In: The California Gold Rush Experience*. New York: Simon & Schuster.

———. 1999. *Rush to Riches: Gold Fever and the Making of California*. Berkeley: Oakland Museum of California and University of California Press.

Holmes, K. L., ed. 1983. *Covered Wagon Women: Diaries and Letters from the Western Trails, 1840–1860*. Vols. 1, 2, 3, 4. Glendale, CA: Arthur H. Clarke Co.

Hunt, T. H. 1974. *Ghost Trails to California*. Las Vegas: Nevada Publications.

King, G. Q. 2003. "Crossing the Forty-Mile Desert." *Overland Journal* 21, no. 4: 122–37.

Levy, J. 1990. *They Saw the Elephant: Women in the California Gold Rush*. Hamden, CT: Shoe String Press.

Marshall, James W. 1857. "To J. M. Hutchings, on the Discovery of Gold January 24, 1848." *Hutchings California Magazine* 2, no. 5:199–201.

Martin, C. W., Jr. 1985. "Geology and the Emigrant: Part I." *Overland Journal* 3, no. 1:4–8; "Geology and the Emigrant: Part II." *Overland Journal* 3, no. 2:17–21; "Geology and the Emigrant: Part III." *Overland Journal* 3, no. 3:28–32.

Mattes, M. J. 1969. *The Great Platte River Road*. Lincoln: University of Nebraska Press.

McCall, A. J. 1882. *The Great California Trail in 1849: Wayside Notes of an Argonaut*. Bath, NY: Steuben Courier. Reprinted from the *Steuben Courier*.

———. 1883. *Pick and Pan: Trip to the Diggins in 1849: Reminiscences of California Life by an Argonaut*. Bath, NY: Steuben Courier. Reprinted from the *Steuben Courier*.

McKinstry, B. 1975. *The California Gold Rush Overland Diary of Byron N. McKinstry, 1850–1852*. Glendale, CA: Arthur H. Clark Co.

McLynn, F. 2002. *Wagons West: The Epic Story of America's Overland Trails*. New York: Grove Press.

Mitchell, J. G. 2000. "The Way West." *National Geographic Magazine*, September 2000, 34–63.

Moeller, B., and J. Moeller. 2001. *The Oregon Trail: A Photographic Journey*. Missoula, MT: Mountain Press Publishing Co.

Morgan, D. E. 1959. *The Overland Diary of James A. Pritchard from Kentucky to California in 1849*. Denver: F. A. Rosenstock.

Munkers, R. L. 1989. "Devils Gate." *Overland Journal* 7, no. 1:2–18.

Myres, S. L. 1986. "I Too Have Seen the Elephant: Women on the Overland Trails." *Overland Journal* 4, no. 4:26–33.

National Frontier Trails Museum. 2004. "Voices from the Trails: Selected Quotes from the National Frontier Trails Museum." http://www.ci.independence.mo.us/NFTM/downloads/quotesbk.pdf.

National Geographic Society. 2005. *National Geographic Historical Atlas of the United States*. Washington, DC: National Geographic Society.

Natrona County Historical Preservation Commission and Rosenberg Historical Consultants. 2001. "Tour Guide: National Historic Trails in Natrona County, Wyoming." Casper, WY: Mountain States Lithographing.

Olch, P. D. 1988. "Treading the Elephant's Tail: Medical Problems on the Overland Trails." *Overland Journal* 6, no. 1:25–31.

Parke, Charles Ross. 1989. *Dreams to Dust: A Diary of the California Gold Rush, 1849–1850*, edited by J. E. Davis. Lincoln: University of Nebraska Press.

Potter, D. M., ed. 1945. *Trail to California: The Overland Journal of Vincent Geiger and Wakeman Bryarly*. New Haven, CT: Yale University Press.

Read, G. W., and R. Gaines. 1949. *Gold Rush: The Journals, Drawings, and Other Papers of J. Goldsborough Bruff*. New York: Columbia University Press.

Rieck, R. L. 1993. "Geography of the California Trails: Part I." *Overland Journal* 11, no. 4:12–22.

———. 1994. "Geography of the California Trails: Part II." *Overland Journal* 12, no. 1:27–32.

———. 1995. "Geography of the California Trails: Part III." *Overland Journal* 13, no. 3:25–32.

Rohrbough, M. J. 1997. *Days of Gold: The California Gold Rush and the American Nation*. Berkeley: University of California Press.

Royce, S. 1932. *A Frontier Lady: Recollections of the Gold Rush and Early California*. New Haven, CT: Yale University Press.

Scamehorn, H. L. 1965. *The Buckeye Rovers in the Gold Rush: An Edition of Two Diaries*. (Contains 1849 diary of John Edwin Banks.) Athens: Ohio University Press.

Schlissel, L. 1982. *Women's Diaries of the Westward Journey*. New York: Schocken Books.

Stewart, G. R. 1962. *The California Trail: An Epic with Many Heroes*. Lincoln: University of Nebraska Press.

Tortorich, F., Jr. 2002. *Gold Rush Trail: A Guide to the Carson River Route of the Emigrant Trail*. Pine Grove, CA: Wagon Wheel Tours.

Unruh, J. D., Jr. 1979. *The Plains Across: The Overland Emigrants and the Trans-Mississippi West, 1840–1860*. Urbana: University of Illinois Press.

Viola, H. J. 1987. *Exploring the West*. Washington, DC: Smithsonian Books.

West, E. 1998. *The Contested Plains*. Lawrence: University Press of Kansas.

Williams, J. 2000. "Water Wasn't Everywhere." *Overland Journal* 18, no. 1:25–28.

Willoughby, R. J. 2003. *The Great Western Migration to the Gold Fields of California, 1849–1850*. Jefferson, NC: McFarland & Co.

GEOLOGY

Adams, K. D., and S. G. Wesnousky. 1998. "Shoreline Processes and the Age of the Lake Lahontan Highstand in the Jessup Embayment, Nevada." *Geological Society of America Bulletin* 110, no. 10:1318–32.

Alt, D., and D. W. Hyndman. 1989. *Roadside Geology of Idaho*. Missoula, MT: Mountain Press.

———. 2000. *Roadside Geology of Northern and Central California*. Missoula, MT: Mountain Press.

Ambrose, P. D. 2004. *Along the Shores of Time*. Vernal, UT: Dinosaur Nature Association in cooperation with Fossil Butte National Monument.

Anderson, D. L. 2000. "The Thermal State of the Upper Mantle; No Role for Mantle Plumes." *Geophysical Research Letters* 27, no. 22:3623–26.

Baldridge, W. S. 2004. *Geology of the American Southwest*. Cambridge: Cambridge University Press.

Blackstone, D. L., Jr. 1988. *Traveler's Guide to the Geology of Wyoming.* 2nd ed. Geological Survey of Wyoming Bulletin 67. Laramie: University of Wyoming.

Bond, J. G., and C. H. Wood. 1978. *Geologic Map of Idaho, Scale 1:500,000.* Moscow, ID: Idaho Department of Lands, Bureau of Mines and Geology.

Boyd, O. S., C. H. Jones, and A. F. Sheehan. 2004. "Foundered Lithosphere Imaged Beneath the Southern Sierra Nevada, USA." *Science* 305 (July 30): 660–62.

Brown, W. G. 1993. "Structural Style of Laramide Basement-Cored Uplifts and Associated Folds." In *Geology of Wyoming,* edited by Arthur W. Snoke, James R. Steidtmann, and Sheila M. Roberts, 312–71. Laramie: Geological Survey of Wyoming Memoir number 5.

Bunge, H.-P., and S. Grand. 2000. "*Mesozoic Plate-Motion History Below the Northeast Pacific Ocean from Seismic Images of the Subducted Farallon Slab.*" *Nature* 405:337–40.

Burchett, R. R. 1986. *Geologic Bedrock Map of Nebraska, Scale 1:1,000,000.* Lincoln: University of Nebraska Conservation and Survey Division, Institute of Agriculture and Natural Resources.

Cannon, W. F. 1960. "The Uniformitarian-Catastrophist Debate." *Isis* 51, no. 1 (March): 38–55.

Carruthers, M. W. 1999. "Hutton's Unconformity: James Hutton's Ideas about Geology." *Natural History Magazine* 108, no. 5 (June): 86–87.

Christiansen, R. L., G. R. Foulger, and J. R. Evans. 2002. "Upper-Mantle Origin of the Yellowstone Hotspot." *Bulletin of the Geological Society of America* 114:1245–56.

Colgan, J. P., T. A. Dumitru, and E. L. Miller. 2004. "Diachroneity of Basin and Range Extension and Yellowstone Hotspot Volcanism in Northwestern Nevada." *Geology* 32:121–24.

Crowley, B., P. Koch, and E. Davis. 2004. "Dating Sierra Uplift with Isotopic Records from Tertiary Mammals." *Journal of Vertebrate Paleontology* 24 (3rd supplement): 49A.

Cutler, A. 2003. *The Seashell on the Mountaintop.* New York: Dutton.

Dalton, R. 2003. "A Window on the Inner Earth." *Nature* 421 (January 2): 10–12.

DeCourten, F. L. 2003. *The Broken Land: Adventures in Great Basin Geology.* Salt Lake City: University of Utah Press.

Dickinson, W. R. 2001. "The Coming of Plate Tectonics to the Pacific Rim." In *Plate Tectonics: An Insider's History of the Modern Theory of the Earth,* edited by N. Oreskes, 264–87. Boulder, CO: Westview.

Dickinson, W. R., M. A. Klute, M. J. Hayes, S. U. Janecke, E. R. Lundin, M. A. McKittrick, and M. D. Olivares. 1988. "Paleogeographic and Paleotectonic Setting of Laramide Sedimentary Basins in the Central Rocky Mountain Region." *Geological Society of America Bulletin* 100:1023–39.

Diffendal, R. F., Jr. 1987. "Ash Hollow State Historical Park: Type Area for the Ash Hollow Formation (Miocene), Western Nebraska." In *Geological Society of America Centennial Field Guide: North-Central Section,* 29–34. Boulder, CO: Geological Society of America.

English. J. M., S. T. Johnston, and K. Wang. 2003. "Thermal Modeling of the Laramide Orogeny: Testing the Flat-Slab Subduction Hypothesis." *Earth and Planetary Sciences Letters* 214:619–32.

Epis, R. C., and C. E. Chapin. 1975. "Geomorphic and Tectonic Implications of the Post-Laramide, Late Eocene Erosion Surface in the Southern Rocky Mountains." *Geological Society of America Memoir* 144:45–74.

Flanagan, K. M., and J. Montagne. 1993. "Neogene Stratigraphy and Tectonics of Wyoming." In *Geology of Wyoming*, edited by A. W. Snoke, J. R. Steidtmann, and S. M. Roberts, 572–607. Laramie: Geological Survey of Wyoming Memoir number 5.

Flesch, L. M., W. E. Holt, A. J. Haines, and B. Shen-Tu. 2000. "Dynamics of the Pacific-North American Plate Boundary in the Western United States." *Science* 287:834–36.

Foulger, G. R., J. G. Fitton, D. C. Presnall, and W. J. Morgan. 2005. "Plumes, Plates, and Paradigms." *Geological Society of America*, Special Volume 388.

Fromm, R., G. Zandt, and S. L. Beck. 2004. "Crustal Thickness Beneath the Andes and Sierras Pampeanas at 30 Degrees S Inferred from Pn Apparent Phase Velocities." *Geophysical Research Letters* 31.

Gilbert, H. J., and A. F. Sheehan. 2004. "Images of Crustal Variations in the Intermountain West." *Journal of Geophysical Research* 109: B03306.

Gould, S. J. 1987. *Time's Arrow, Time's Cycle*. Cambridge, MA: Harvard University Press.

Grand, S. P., R. D. van der Hilst, and S. Widiyantoro. 1997. "Global Seismic Tomography: A Snapshot of Convection in the Earth," *GSA Today* 7, no. 4:1–7.

Gregory, K. M., and C. G. Chase. 1992. "Tectonic Significance of Paleobotanically Estimated Climate and Altitude of the Late Eocene Erosion Surface, Colorado." *Geology* 20:581–85.

Hamblin, W. K., and E. H. Christiansen. 2004. *Earth's Dynamic Systems*. 10th ed. Upper Saddle River, NJ: Prentice-Hall.

Hamblin, W. K., J. K. Rigby, J. L. Snyder, and W. H. Matthews. 1974. *Roadside Geology of U.S. Interstate 80 between Salt Lake City and San Francisco*. Van Nuys, CA: Varna Enterprises.

Harden, D. R. 2004. *California Geology*. 2nd ed. Upper Saddle River, NJ: Pearson Prentice Hall.

Hill, M. 2002. *Gold: The California Story*. Berkeley: University of California Press.

Hoffman, P. F. 1988. "United Plates of America, the Birth of a Craton: Early Proterozoic Assembly and Growth of Laurentia." *Annual Reviews of Earth and Planetary Sciences* 16:543–603.

Houston, R. S. 1993. "Late Archean and Early Proterozoic Geology of Southeastern Wyoming." In *Geology of Wyoming*, edited by A. W. Snoke, J. R. Steidtmann, and S. M. Roberts, 78–116. Laramie: Geological Survey of Wyoming Memoir number 5.

Humphreys, E. D. 1995. "Post-Laramide Removal of the Farallon Slab, Western United States." *Geology* 25:987–90.

Humphreys, E. D., K. G. Dueker, D. L. Schutt, and R. B. Smith. 2000. "Beneath Yellowstone: Evaluating Plume and Nonplume Models Using Teleseismic Images of the Upper Mantle." *GSA Today* 10, no. 12 (December): 1–7.

Hutton, J. 1785. "Abstract of a dissertation read in the Royal Society of Edinburgh upon the seventh of March and fourth of April 1785, concerning the system of the Earth, its duration and stability." Reprinted in *Philosophy of Geohistory, 1785–1970*, edited by C. C. Albritton, 24–52. Stroudsburg, PA: Dowden, Hutchinson & Ross, 1975.

———. 1788. "Theory of the Earth; or, An Investigation of the Laws Observable in the Composition, Dissolution, and Restoration of Land upon the Globe." *Transactions of the Royal Society of Edinburgh* 1:209–304. (This is the text of the paper Hutton read on March 7 and April 4, 1785, before the Royal Society of Edinburgh.)

———. 1795. *Theory of the Earth, with Proofs and Illustrations.* 2 vols. Edinburgh: William Creach. Reprinted, New York: J. Cramer, 1972.

Jahren, A. H. 2002. "The Biogeochemical Consequences of the Mid-Cretaceous Superplume." *Journal of Geodynamics* 34:177-91.

Jones, N. 2003. "Volcanic Bombshell." *New Scientist* 177 (March 8): 32-35.

Kay, S. M., and C. Mpodozis. 2001. "Central Andean Ore Deposits Linked to Evolving Shallow Subduction Systems and Thickening Crust." *GSA Today*, March, 4-9.

Kent, G. M, et al. 2005. "60 K.Y. Record of Extension Across the Western Boundary of the Basin and Range Province: Estimate of Slip Rates from Offset Shoreline Terraces and a Catastrophic Slide Beneath Lake Tahoe." *Geology* 33, no. 5:365-68.

Kirkemo, H., W. L. Newman, and R. P. Ashley. 2006. *Gold.* Denver: U.S. Geological Survey.

Lageson, D. R., and D. R. Spearing. 1991. *Roadside Geology of Wyoming.* 2nd revised ed. Missoula, MT: Mountain Press.

Larson, R. L. 1991. "Latest Pulse of Earth: Evidence for a Mid-Cretaceous Superplume." *Geology* 19:547-50.

———. 1995. The Mid-Cretaceous Superplume Episode." *Scientific American* 272 (February): 66-70.

Love, D. L., and A. C. Christiansen. 1985. *Geologic Map of Wyoming, 1:500,000.* Denver: Department of the Interior, U.S. Geological Survey.

Lowenstern, J. B. 2005. "Truth, Fiction, and Everything in Between at Yellowstone." *Geotimes* 50, no. 6 (June): 18-23.

Lowenstern, J. B., R. L. Christiansen, R. B. Smith, L. A. Morgan, and H. Heasler. 2005. "Steam Explosions, Earthquakes, and Volcanic Eruptions—What's in Yellowstone's Future?" United Stated Geological Survey Fact Sheet 2005-3024.

Lyell, Charles. 1830. *The Principles of Geology.* 1st ed. 3 vols. London: J. Murray. Reprint, Chicago: University of Chicago Press, 1990.

Maher, H. D., Jr., G. F. Englemann, and R. D. Shuster. 2003. *Roadside Geology of Nebraska.* Missoula, MT: Mountain Press.

Marshak, S. 2001. *Earth: Portrait of a Planet.* New York: W. W. Norton.

Martin, C. W., Jr. 1985. "Geology and the Emigrant: Part I." *Overland Journal* 3, no. 1:4-8; "Geology and the Emigrant: Part II." *Overland Journal* 3, no. 2:17-21; "Geology and the Emigrant: Part III." *Overland Journal* 3, no. 3:28-32.

———. 1989. "Geology of Silent City of Rocks." *Overland Journal* 7, no. 4:24-27.

McDonald, R. E. 1972. "Eocene and Paleocene Rocks of the Southern and Central Basins." In *Geologic Atlas of the Rocky Mountain Region,* edited by W. W. Mallory, 243-56. Denver: Rocky Mountain Association of Geologists, A.B. Hirschfeld Press.

McMillan, M. E., C. L. Angevine, and P. L. Heller. 2002. "Postdepositional Tilt of the Miocene-Pliocene Ogallala Group on the Western Great Plains: Evidence of Late Cenozoic Uplift of the Rocky Mountains." *Geology* 30:63-66.

McMillan, M.E., P. L. Heller, and S. L. Wing. 2006. "History and Causes of Post-Laramide Relief in the Rocky Mountain Orogenic Plateau." *Geological Society of America Bulletin* 118, nos. 3/4:393-405.

McPhee, John. 1998. *Annals of the Former World.* New York: Farrar, Straus and Giroux.

Mears, B., Jr. 1993. "Geomorphic History of Wyoming and High Level Erosion Surfaces." In *Geology of Wyoming,* edited by A. W. Snoke, J. R. Steidtmann, and S. M. Roberts, 608-26. Laramie: Geological Survey of Wyoming Memoir number 5.

Molnar, P., and P. England. 1990. "Late Cenozoic Uplift of Mountain Ranges and Global Climate Change: Chicken or Egg?" *Nature* 346:29-34.

Montelli, R., G. Nolet, F. A. Dahlen, G. Masters, E. R. Engdahl, and S.-H. Hung. 2004. "Finite-Frequency Tomography Reveals a Variety of Plumes in the Mantle." *Science* 303 (January 16): 338-43.

Morgan, W. J. 1971. "Convection Plumes in the Lower Mantle." *Nature* 230:42-43.

Mosel, S. S. 2004. "From Water Hole to Rhino Barn." *Natural History,* September 2004, 56-57.

Mulch, A., S. A. Graham, and C. P. Chamberlain. 2006. "Hydrogen Isotopes in Eocene River Gravels and Paleoelevation of the Sierra Nevada." *Science* 313 (July 7): 87-89.

Muntean, J. 2006. "The Rush to Uncover Gold's Origins." *Geotimes,* April 2006, 24-27.

Oldow, J. S., A. W. Bally, H. G. Avé Lallemant, and W. P. Leeman. 1989. "Phanerozoic Evolution of the North American Cordillera; United States and Canada." In *The Geology of North America: An Overview.* Geology of North America Volume A, edited by A. W. Bally and A. R. Palmer, 139-232. Boulder, CO: Geological Society of America.

Orndorff, R. L., R. W. Wieder, and H. F. Filkorn. 2001. "Finding Fault—the Genoa Fault Scarp." In *Geology Underfoot in Central Nevada,* by R. L. Orndorff, R. W. Wieder, and H. F. Filkorn, 38-47. Missoula, MT: Mountain Press.

Playfair, John. 1802. *Illustrations of the Huttonian Theory of the Earth.* Edinburgh: William Creach. Reprint, New York: Dover, 1964.

Powell, John Wesley. 1875. *Exploration of the Colorado River of the West and Its Tributaries.* Washington, DC: U.S. Government Printing Office.

Rehels, M. 1999. *Extent of Pleistocene Lakes in the Western Great Basin.* U.S. Geological Survey, Miscellaneous Field Studies Map 2323.

Repcheck, Jack. 2003. *The Man Who Found Time.* Cambridge, MA: Perseus.

Royse, F., Jr. 1993. "An Overview of the Geologic Structure of the Thrust Belt in Wyoming, Northern Utah and Eastern Idaho." In *Geology of Wyoming,* edited by A. W. Snoke, J. R. Steidtmann, and S. M. Roberts, 272-311. Laramie: Geological Survey of Wyoming Memoir number 5.

Royse, F., Jr., and M. A. Warner. 1987. "Little Muddy Creek Area, Lincoln County, Wyoming." In *Geological Society of America Centennial Field Guide—Rocky Mountain Section,* 213-16. Boulder, CO: Geological Society of America.

Sahagian, D., A. Proussevitch, and W. Carlson, W. 2002. "Timing of Colorado Plateau Uplift: Initial Constraints from Vesicular Basalt-Derived Paleoelevations." *Geology* 30:807-10.

Saleeby, J. 2003. "Segmentation of the Laramide Slab: Evidence from the Southern Sierra Nevada." *Geological Society of America Bulletin* 115 (June): 655-68.

Saleeby, J., and Z. Foster. 2004. "Topographic Response of Mantle Lithosphere Removal in the Southern Sierra Nevada Region, California." *Geology* 32, no. 3:245-48.

Sayles, J. K. 1983. "Collapse of Rocky Mountain Basement Uplifts." In *Rocky Mountain Foreland Basins and Uplifts,* edited by J. D. Lowell, 79-97. Denver: Rocky Mountain Association of Geologists.

Sharp, R. P., and A. F. Glazner. 1997a. "A Frightful Earthquake: The Owens Valley Shock of 1872." In *Geology Underfoot in Death Valley and Owens Valley,* by R. P. Sharp and A. F. Glazner, 195-201. Missoula, MT: Mountain Press.

———. 1997b. "A Lunar Landscape: The Tufa Pinnacles of Searles Lake." In *Geology Underfoot in Death Valley and Owens Valley,* by R. P. Sharp and A. F. Glazner, 33–40. Missoula, MT: Mountain Press.

Silva, M. 1986. *Placer Gold Recovery Methods.* Sacramento: California Department of Conservation, Division of Mines and Geology Special Publication 87.

Slemmons, D. B., and J. W. Bell. 1987. "The 1954 Fairview Peak Earthquake Area, Nevada." In *Geological Society of America Centennial Field Guide—Cordilleran Section,* 73–76. Boulder, CO: Geological Society of America.

Smith, R. B., and L. J. Siegel. 2000. *Windows into the Earth: The Geologic Story of Yellowstone and Grand Teton National Parks.* Oxford: Oxford University Press.

Snoke, A. W. 1993. "Geologic History of Wyoming within the Tectonic Framework of the North American Cordillera." In *Geology of Wyoming,* edited by A. W. Snoke, J. R. Steidtmann, and S. M. Roberts, 2–56. Laramie: Geological Survey of Wyoming Memoir number 5.

Stanley, Steven M. 2005. *Earth System History,* 2nd ed. New York: W. H. Freeman.

Steidtmann, J. R. 1993. "The Cretaceous Foreland Basin and Its Sedimentary Record." In *Geology of Wyoming,* edited by A. W. Snoke, J. R. Steidtmann, and S. M. Roberts, 250–71. Laramie: Geological Survey of Wyoming Memoir number 5.

Steinberger, B., R. Sutherland, and R. J. O'Connell. 2004. "Prediction of Emperor-Hawaii Seamount Locations from a Revised Model of Global Plate Motion and Mantle Flow." *Nature* 430 (July 8): 167–73.

Stewart, J. H., and J. E. Carlson. 1978. *Geologic Map of Nevada, Scale 1:500,000.* Denver: U.S. Geological Survey.

Stock, G. M., R. S. Anderson, and R. C. Finkel. 2004. "Pace of Landscape Evolution in the Sierra Nevada, California, Revealed by Cosmogenic Dating of Cave Sediments." *Geology* 32, no. 3:193–96.

Swinehart, J. B., and D. B. Loope. 1987. "Late Cenozoic Geology Along the Summit to Museum Hiking Trail, Scotts Bluff National Monument, Western Nebraska." In *Geological Society of America Centennial Field Guide—North-Central Section,* 13–18. Boulder, CO: Geological Society of America.

Tarduno, J. A., et al. 2003. "The Emperor Seamounts: Southward Motion of the Hawaiian Hotspot Plume in Earth's Mantle." *Science* 301 (August 23): 1064–69.

Tyson, N. deGrasse, C. Liu, and R. Irion. 2000. *One Universe.* Washington, DC: Joseph Henry Press.

Unruh, J., J. Humphrey, and A. Barron, A. 2003. "Transtensional Model for the Sierra Nevada Frontal Fault System, Eastern California." *Geology* 33, no. 4:327–30.

Wakabayashi, J., and G. M. Stock. 2003. "Overview of the Cenozoic Geologic History of the Sierra Nevada." In *Tectonics, Climate Change, and Landscape Evolution in the Southern Sierra Nevada, California: Pacific Cell Friends of the Pleistocene Field Trip Guidebook,* edited by G. M. Stock, 31–40. Santa Cruz: University of California, Santa Cruz.

Wernicke, B., and J. K. Snow. 1998. "Cenozoic Tectonism in the Central Basin and Range: Motion of the Sierra Nevada—Great Valley Block." *International Geology Review* 40, no. 5:403–10.

Wolfe, J. A., C. E. Forest, and P. Molnar. 1998. "Paleobotanical Evidence of Eocene and Oligocene Paleoaltitudes in Midlatitude Western North America." *Geological Society of America Bulletin* 110, no. 5:664–78.

Wolfe, J. A., H. E. Schorn, C. E. Forest, and P. Molnar. 1997. "Paleobotanical Evidence for High Altitudes in Nevada During the Miocene." *Science* 276:1672–75.

Wyckoff, R. M. 1999. *Hydraulicking North Bloomfield and the Malakoff Diggins State Historic Park*. Nevada City, CA: Robert M. Wyckoff.

Yuan, H., and K. Dueker. 2005. "Teleseismic P-Wave Tomogram of the Yellowstone Plume." *Geophysical Research Letters* 32, no. 7.

Zandt, G., H. Gilbert, T. J. Owens, M. Ducea, J. Saleeby, and C. H. Jones. 2004. "Active Foundering of a Continental Arc Root Beneath the Southern Sierra Nevada in California." *Nature* 431 (September 2): 41–45.

FIGURE CREDITS

All shaded relief bases on shaded relief maps are courtesy of the U.S. Geological Survey EROS Data Center, Sioux Falls, South Dakota. All photographs and artwork not credited below are by the author.

Introduction photograph of unknown California miner: Pick Pan & Shovel, ca. 1850, reproduced by permission, Collection of Matthew R. Isenburg.

Figure 2.1: A Pioneer Family on the Great Plains (date uncertain), Denver Public Library, Western History Collection, X-11929.

Figure 2.2: Seeing the Elephant, 1850, California Historical Society, FN-04479.

Figure 2.5: James Hutton portrait by Abner Lowe, Museum Property, U.S. Geological Survey. Charles Lyell portrait from a daguerreotype by J. E. Mayal, reproduced in Arthur Shuster and Arthur E. Shipley, *Britain's Heritage of Science* (London: Constable & Co. Ltd., 1917), facing p. 310.

Figure 3.1: Aerial photograph of Platte River Valley courtesy U.S. Geological Survey.

Figure 3.2: J. Goldsborough Bruff drawing "Cooking on the Plains" reproduced by permission of the Huntington Library, San Marino, California.

Figure 3.3: J. Goldsborough Bruff drawing "A camp-scene" reproduced by permission of the Huntington Library, San Marino, California.

Figure 3.4: Based on DeCourten 2003, fig. 2.4, p. 28.

Figures 3.5 and 3.6: Based on Hamblin and Christiansen 2004, fig. 21.30, p. 628; Snoke 1993, fig. 4, p. 8; and Baldridge 2003, fig. 1.4, p. 14.

Figure 4.3: J. Goldsborough Bruff drawing "Tremendous hailstorm" reproduced by permission of the Huntington Library, San Marino, California.

Figure 4.7: Modified from Blackstone 1988, fig. 67, p. 117, reproduced by permission of the Wyoming State Geological Survey.

Figure 5.3: Photograph of Bighorn River through Sheep Mountain reproduced by permission of Peter Huntoon.

Figure 5.4: Based on Lageson and Spearing 1991, 20.

Figure 5.6: Modified from Mears 1993, figs. 2, 3 4, 6, 7, pp. 613-14, reproduced by permission of the Wyoming State Geological Survey.

Figure 5.7: Photograph of the Wind River Range subsummit surface reproduced by permission of David Lageson.

Figure 5.8: J. Goldsborough Bruff drawing "Ferriage of the Platte" reproduced by permission of the Huntington Library, San Marino, California.

Figure 6.4: J. Goldsborough Bruff drawing "A View of the Sweet Water River" reproduced by permission of the Huntington Library, San Marino, California.

Figure 6.7: Based on Flanagan and Montagne 1993, figs. 7, 8, pp. 584-85; and Lageson and Spearing 1991, 125.

Figures 7.1-7.4 include portions of digital art supplied to the public realm by the U.S. Geological Survey.

Figure 7.5: Virtual image of South America reproduced by permission of William Bowen.

Figure 7.10: Modified from Blackstone 1988, fig. 34, p. 49, reproduced by permission of the Wyoming State Geological Survey.

Figure 8.2: J. Goldsborough Bruff drawing "Great Dividing Ridge of the Rocky Mountains" reproduced by permission of the Huntington Library, San Marino, California.

Figure 8.3: J. Goldsborough Bruff drawing "Terminus of the Greenwood cut-off" reproduced by permission of the Huntington Library, San Marino, California.

Figure 8.7: Modified from DeCourten 2003, fig. 6.21, p. 142, reproduced by permission of the University of Utah Press.

Figure 9.4: J. Goldsborough Bruff drawing "Steam Boat Spring" reproduced by permission of the Huntington Library, San Marino, California.

Figure 9.6: Based on Smith and Siegel 200, fig. 1.3, p. 10, and fig. 3.1, p. 48.

Figure 9.7: Modified from DeCourten 2003, fig. 8.10, p. 178, reproduced by permission of the University of Utah Press.

Figure 10.7: Based on DeCourten 2003, fig. 8.17, p. 185.

Figure 10.9: Based on Flesch et al. 2000, fig. 1, p. 834.

Figure 11.4: Lower photograph of entrenched meanders in Palisade Canyon courtesy of the U.S. Geological Survey.

Figure 11.6: Aerial photograph of Humboldt Valley courtesy of the U.S. Geological Survey.

Figure 11.7: J. Goldsborough Bruff drawing "The Rabbit-Hole Springs" reproduced by permission of the Huntington Library, San Marino, California.

Figure 12.4: Based on the map of Rehels 1999.

Figure 12.10: Based on Wernicke and Snow 1998, fig. 2, p. 406.

Figure 13.3: Based on Stewart 1962, p. 117.

Figure 13.6: Based on Harden 2004, fig. 8.37, p. 198, with additional data from the California Geological Survey.

Figure 13.7: Photograph by Carleton E. Watkins, "Malakoff Diggins, North Bloomfield, Nevada County, California," ca. 1869. San Francisco Museum of Modern Art, purchased through a gift of the Judy Kay Memorial Fund, SFMOMA internal number 96.125. Inset photograph (by author) of monitor is from Marshall Gold Discovery State Park.

Figure 13.9: Based on Harden 2004, fig. 8.30, p. 190, with additional data supplied by the California Geological Survey.

Epilogue photograph: 1869 photograph of Jupiter Train used by permission, Utah State Historical Society, all rights reserved.

INDEX

*Pages with illustrations are referred to
by the page number followed by f.*